Management for Professionals

The Springer series "Management for Professionals" comprises high-level business and management books for executives, MBA students, and practice-oriented business researchers. The topics cover all themes relevant to businesses and the business ecosystem. The authors are experienced business professionals and renowned professors who combine scientific backgrounds, best practices, and entrepreneurial vision to provide powerful insights into achieving business excellence.

The Series is SCOPUS-indexed.

Konstantina Ragazou
Constantin Zopounidis • Emilios Galariotis
Nikolaos Sariannidis • Georgia Zournatzidou

Transparency and Corporate Washing in the Age of Sustainability

An Operational Research Agenda to Evaluate Corporate Sustainability Disclosure

Springer

Konstantina Ragazou
Department of Management Science and Technology
University of Western Macedonia
Kozani, Greece

Constantin Zopounidis
School of Production Engineering and Management
Technical University of Crete
Chania, Greece

Emilios Galariotis
Bang College of Business
KIMEP University
Almaty, Kazakhstan

Nikolaos Sariannidis
Department of Accounting and Finance
University of Western Macedonia
Kozani, Greece

Georgia Zournatzidou
Department of Business Administration
University of Western Macedonia
Grevena, Greece

ISSN 2192-8096 ISSN 2192-810X (electronic)
Management for Professionals
ISBN 978-3-031-96820-4 ISBN 978-3-031-96821-1 (eBook)
https://doi.org/10.1007/978-3-031-96821-1

This Springer imprint is published by the registered company Springer Nature Switzerland AG
The registered company address is: Gewerbestrasse 11, 6330 Cham, Switzerland

Preface

ESG investors are individuals who conduct a comprehensive analysis of a company's ecological/environmental, social, and green components prior to making an investment. This is a widely recognized definition of ESG investors. It is reasonable to expect that a greater number of individuals will have the desire to invest in or operate their enterprises in a sustainable and environmentally favorable manner as time progresses. This is due to the fact that climate change is becoming a more substantial threat to humanity. The increasing prevalence of environmental, social, and governance (ESG) policies has been observed by both large and small organizations. This is attributable to the fact that sustainability has become the new motto that is increasingly frequently heard on a global scale. In 2015, the United Nations established a list of sustainable development objectives for the first time. Subsequently, an increasing number of nations have integrated these objectives into their distinctive growth strategies.

Consequently, the objective of this book is to investigate the growing emphasis on sustainable development, which once more reveals the unethical business practices that companies employ to increase their profits. The majority of organizations have finally acknowledged that they cannot continue to disregard the increasing demand for sustainability. As a result, they have initiated the process of modifying their operations to reflect this realization. Greenwashing is a controversial strategy that has been employed by numerous organizations to create the false impression that their operations are environmentally friendly. Nevertheless, numerous corporations persist in employing environmentally harmful methods and processes in their operations. In the subsequent paragraphs, we will examine the numerous forms of greenwashing, bluewashing, social washing, and pinkwashing that are employed by companies.

Kozani, Greece — Konstantina Ragazou
Chania, Greece — Constantin Zopounidis
Almaty, Kazakhstan — Emilios Galariotis
Kozani, Greece — Nikolaos Sariannidis
Grevena, Greece — Georgia Zournatzidou

Contents

1 Greenwashing: The Misleading Strategies Underlying Environmental Claims

1.1 Defining Greenwashing

The existing literature presents a divergence of opinions regarding the definition of greenwashing. In general, greenwashing can be defined as the practice of deceiving individuals into forming favorable perceptions regarding the environmental performance of an organization or its products and services. Greenwashing manifests in various forms, encompassing the selective application of ambiguous sustainability assertions to blatant falsehoods regarding a product's or an organization's sustainability performance. Greenwashing is not an inadvertent mistake; rather, it constitutes a calculated effort to deceive stakeholders. The Volkswagen scandal, characterized by the manufacturer's manipulation of emission data for its vehicles, serves as a prominent illustration of corporate misconduct.

Throughout the years, various specific forms of greenwashing have emerged, including CSR (corporate social responsibility)-washing, blue-washing, and SDG (sustainable development goal)-washing. Moreover, overlapping concepts such as "window-dressing" and "smoke-screening" are occasionally employed as synonyms for "greenwashing" to characterize the selective disclosure of favorable environmental information, which does not accurately reflect the true environmental performance of a product or company (Li, 2024; Zhang, 2022). Finally, academicians have developed terminology to characterize the remedy for greenwashing, a phenomenon in which a company deliberately minimizes the communication of its environmental practices. For example, the notion of brownwashing refers to the practice of obscuring the expenses associated with sustainability initiatives through an exaggerated sense of humility. Greenwashing and strategic silence are two other terms used to describe companies that underreport rather than overreport their environmental performance (Fella & Bausa, 2024).

Greenwashing encompasses a diverse array of academic disciplines. Theoretically, various additional concepts also pertain to the discrepancy between organizational

K. Ragazou et al., *Transparency and Corporate Washing in the Age of Sustainability*, Management for Professionals,
https://doi.org/10.1007/978-3-031-96821-1_1

discourse and actual conduct. Numerous prior studies have investigated the concept of decoupling, which was initially employed to elucidate how organizations safeguard their fundamental operations from external expectations by symbolically implementing 'inefficient' policies and structures. In this context, "decoupling" denotes a disjunction between policy formulation and its execution, thereby enabling the organization to sustain its daily operations while simultaneously addressing a multitude of external expectations (Tzitiridou-Chatzopoulou & Zournatzidou, 2024; Zournatzidou, 2024; Zournatzidou & Floros, 2023). Decoupling has also been employed in the examination of sustainability and greenwashing, particularly in instances where oil companies redirect their focus from crises and environmentally detrimental business practices. Academics have also examined the phenomenon of means-ends decoupling, which transpires when there exists a degree of consistency between articulated intentions and actual actions; however, the latter is largely unrelated to the organization's fundamental objectives and processes. Fast fashion enterprises, for instance, implement means-ends decoupling by segregating their circular economy initiatives from their fundamental business operations (Koemtzopoulos et al., 2025; Tzitiridou-Chatzopoulou & Zournatzidou, 2024; Zournatzidou, Sklavos, Ragazou, & Sariannidis, 2024a). Nevertheless, decoupling can be characterized as greenwashing only when an organization deliberately seeks to mislead stakeholders regarding its sustainability practices in order to enhance its reputation, attract resources, or obscure contentious activities. Decoupling may also arise in circumstances where the relationship between discourse and actual behavior is intricate, ambiguous, and dubious, rendering it unclear which actions will enable the organization to fulfill its commitments.

Organizational hypocrisy constitutes an additional area of scholarly inquiry that examines the disjunction between rhetoric and action. Extant literature on hypocrisy assumes that talk-action consistency is difficult to accomplish, for example, due to a lack of knowledge, resources, time, and feasibility (Ragazou et al., 2024; Tzitiridou-Chatzopoulou et al. 2024; Zournatzidou et al. 2024). Organizations face a multitude of stakeholder demands that are often inconsistent, requiring them to address diverse interests through a range of discussions and actions. Moreover, individuals who engage in verbal communication and those who take action may exhibit differences within organizations, leading to discrepancies between spoken statements and actual behaviors. As articulated by Brunsson (1993, p. 496), '(...) individuals who possess the liberty to articulate ideas without the obligation to implement them are frequently in a position to advocate for perspectives that are more ethical, esthetically pleasing, or truthful, albeit less practical, than those that must inform the actions of the actors involved' (Brunsson, 1993). Although the term "hypocrisy" is frequently linked to a deficiency in moral integrity, scholars argue that some degree of inconsistency between verbal commitments and actual behavior is inevitable and constitutes a standard practice within organizations. In certain instances, duplicity may indeed generate opportunities for transformation that would not have been feasible under different circumstances. In conclusion, hypocrisy and greenwashing intersect solely when the inconsistencies between verbal commitments and actual practices are founded upon misleading efforts to deceive stakeholders regarding the organization's sustainability performance.

There exists a modest yet expanding body of literature concerning organizational rhetoric, commonly referred to as "organizational bullshit." This term is defined as corporate discourse that cultivates a favorable perception of an individual or entity, irrespective of the actual circumstances within the organization. Traditionally linked primarily to spoken discourse, the notion of 'bullshit' has recently garnered attention within academic research in the fields of management and organizational studies. A disingenuous company is characterized by its pervasive indifference toward factual accuracy and the authentic circumstances of its operations. As articulated by Spicer (p. 657), '(…) a significant amount of discourse and written communication within organizations appears to be intrinsically "hollow," exhibiting minimal correlation with the actual dynamics occurring within the organization' (Koch & Denner, 2025; Li et al. 2024). Bullshit may be defined as discourse and written communication that exhibits a disregard for veracity. Nonetheless, not all organizational obfuscation is a conscious effort to deceive its stakeholders. For example, corporate terminologies such as 'disruption,' 'resilience,' and 'agility' do not invariably pertain to schemes against others or oneself. Although insincere discourse may undoubtedly serve the interests of the individual engaging in it, certain forms of 'empty talk' and 'heated air' within organizations can also function as socially acceptable conventions that signify membership within a community, irrespective of their veracity. In contrast to an organization engaged in greenwashing, which deliberately deceives stakeholders regarding its environmental practices, a bullshitting organization exhibits a general indifference toward the truth and the actual state of its operations. As articulated by de Freitas Netto et al. (2020), "The fundamental aspect of the bullshitter's character that remains concealed is that the truth-values of his assertions hold no significant relevance for him; what must not be misconstrued is that his objective is neither to convey the truth nor to obscure it." This does not imply that his speech is chaotically impulsive; rather, it suggests that the underlying motive that directs and governs it is indifferent to the actual nature of the subjects he addresses (Brandão et al., 2023; Li et al. 2024; Li & Chen, 2024; Wang et al. 2024). In conclusion, several terminologies within the management and organizational literature pertain to the discrepancies between an organization's communication and its actual practices. Greenwashing is distinct from the aforementioned concepts in that it specifically pertains to environmental matters and underscores intentional endeavors to deceive consumers or other stakeholders. Nonetheless, as articulated by Seele and Gatti (2017), a comprehensive understanding of greenwashing necessitates consideration of the perspectives of those individuals or groups alleging such practices against the organization, including activists, media representatives, consumer advocacy agencies, and others. Similar to how one organization may possess a motivation to engage in greenwashing, another organization may be incentivized to accuse others of such practices. This could manifest as attributing blame to those perceived as insufficiently proactive or failing to adhere to their stated environmental commitments. Consequently, an analysis of greenwashing must take into account both the party being implicated and the party making the accusation.

1.2 Determinants of Greenwashing

In order to understand the motivations behind greenwashing, it is essential to first analyze key determinants that, on the one hand, encourage the occurrence of greenwashing and, on the other hand, influence its containment. The key factors are listed below.

1.2.1 Product Market Competition

Upon examining the competitive dynamics among individual companies within the market, alongside the occurrence of greenwashing, one may initially perceive this competition as a double-edged sword. On the one hand, this phenomenon can be attributed to the fact that a highly competitive market typically engenders a competitive environment among companies. Greenwashing refers to the practice of misleading consumers into believing that an organization is more dedicated to social and environmental responsibilities than it truly is (Wang et al. 2024). This can be particularly advantageous for a CSR-sensitive customer base. In contrast, organizations that engage in transparent corporate social responsibility (CSR) reporting yet implement only minimal social and environmental initiatives within their business operations find themselves at a disadvantage in the current environment. To maintain their competitive edge, the resultant impairment may incentivize these companies to resort to greenwashing practices. It is thus plausible that a chain reaction may be initiated within a highly competitive environment, where greenwashing, as an unethical business practice, becomes prevalent (Zournatzidou et al. 2024a, 2024).

Conversely, enterprises that partake in greenwashing face the potential peril of their inaccurate disclosures being unveiled, which could subsequently result in adverse repercussions for market participants. In a highly competitive market, consumers may, in instances of uncertainty, opt for substitute products and refrain from supporting companies that engage in greenwashing practices. As noted by Arouri et al. (2021), firms that manufacture products with significant environmental repercussions in a competitive marketplace tend to disclose supplementary information regarding their corporate social responsibility (CSR) initiatives (Arouri et al., 2021). This incentive arises from the tendency of companies with a negative corporate social responsibility (CSR) record to leverage publications as a means of enhancing their image and bolstering trust in both themselves and their products. Consequently, consumers may be less inclined to consider alternatives that seem to possess a reduced environmental footprint. In conjunction with this corporate advantage, the supplementary publications also mitigate the information asymmetry that exists between the company and its stakeholders. This phenomenon arises from the efforts of companies to distinguish themselves from their competitors by means of comprehensive and transparent disclosures. Stakeholders are afforded the opportunity to examine and contrast the comprehensive reports, thereby diminishing the potential for companies to engage in greenwashing practices. In contrast, firms operating within a less competitive market possess minimal motivation to disclose supplementary information regarding their social and environmental performance, attributable to the limited availability of substitute products (Tzitiridou-Chatzopoulou et al. 2024; Zournatzidou et al., 2024, 2024b, 2024). This situation engenders an agency conflict as delineated by the principal–agent theory, in which the organization may seek to exploit the information asymmetry through the practice of greenwashing (Teti et al., 2024). Drawing from the studies examined, it can be posited that while competition results in an increased volume of publications, these reports provide a greater degree of detail, thereby allowing stakeholders to identify inconsistencies and seek alternatives as necessary. Consequently, a competitive environment may serve as a constraint on greenwashing under specific conditions (Bottega et al., 2024; Li & Xiao, 2025; Wang & Shen, 2024; Yu et al., 2020).

1.2.2 Personality Traits of CEOs

Personal values, along with psychological characteristics, significantly influence an individual's leadership style, thereby shaping the trajectory of the organization. The Chief Executive Officer (CEO) of an organization is regarded as the preeminent decision-maker within the entity. Consequently, his personal values, experiences, and psychological attributes exert a substantial impact on the company's strategies, actions, and decision-making processes. Concerning corporate social responsibility, the legislature has enacted a limited number of mandatory provisions, indicating that the majority of actions undertaken by the company are implemented on a voluntary basis. In accordance with the Upper Echelon Theory, it is the personal values, experiences, and psychological attributes of the manager that influence the actions of the company, thereby affecting the execution of a Corporate Social Responsibility (CSR) strategy.

Chin et al. (2013) discovered in their research that organizations led by liberal Chief Executive Officers exhibit a more robust Corporate Social Responsibility profile compared to those managed by conservative executives. It has been observed that an increase in the authority of the liberal Chief Executive Officer correlates with a heightened level of Corporate Social Responsibility within the organization. Furthermore, the corporate social responsibility (CSR) levels exhibited by CEOs who espouse liberal values remained elevated, irrespective of the financial condition of the company. In contrast, conventional managers implemented socially and environmentally responsible initiatives solely when the financial circumstances permitted. Irrespective of whether the attitudes of the CEOs were influenced by liberal or conservative ideologies, the research conducted by Chin et al. (2013) unequivocally demonstrated that a significant number of corporate social responsibility (CSR) initiatives are predominantly symbolic in nature. This suggests that the company's management engages in corporate social responsibility initiatives primarily for the purpose of enhancing its public image. A significant transformation toward a socially and environmentally responsible organization typically does not occur under either liberal or conservative management. Conversely, the fundamental objective is inherently self-serving. Consequently, corporate social responsibility initiatives are implemented to enhance the reputation, status, or self-perception of the manager (Fella & Bausa, 2024; Jiang et al., 2024; Liu, 2024; Sun et al., 2024; Tong et al., 2024).

When a Chief Executive Officer employs socially and environmentally responsible initiatives to enhance their external image, such actions are frequently motivated by underlying narcissistic personality traits. Historically, narcissism has been characterized as a personality disorder marked by an overwhelming need for validation, an inflated sense of self-admiration and self-importance, and a notable deficiency in empathy toward others. In contemporary discourse, narcissism is increasingly regarded as a personality trait rather than a clinical disorder. Narcissistic traits manifest within the population to differing extents and, contingent upon their specific nature, significantly affect an individual's interactions with others, self-perception, and ability to navigate challenges. Nonetheless, individuals exhibiting narcissistic traits are distinguished by an aspiration for notoriety and authority, coupled with a self-perception that leads them to regard themselves as more appealing and intellectually superior to their counterparts. This results in narcissists frequently assuming a dominant position, which engenders a sense of superiority over others and enables them to utilize individuals as subordinates. Consequently, it is not unexpected that the literature frequently identifies evidence of narcissistic characteristics within the management teams of corporations.

A narcissistic CEO has the authority to dictate the company's strategic direction and initiatives. Consequently, initiatives that get media attention are often favored, as shown by CSR. Social and environmental performance is highly favored by stakeholders, resulting in increased media visibility for the CEO. The research conducted by Petrenko et al. (2016) corroborates this concept. Thus, the pursuit of attention and adoration compels the firm to engage in CSR, with the CEO being seen as the architect and originator, garnering acclaim for these efforts. The study's findings indicate a favorable correlation between narcissistic traits in CEOs and a company's corporate social responsibility. The research elucidates that narcissistic CEOs use CSR to fulfill their need for attention, disregarding company success Xia et al. (2023). Al-Shammari et al. (2019) assert that the impetus for social and environmental performance is often neither financial nor ethical in character. The psychological demands of the CEO take precedence, facilitating execution. This not only incites a principal–agent dispute between shareholders and the CEO but also contravenes the conventional standard of the respectable businessman. This notion underscores that a respectable businessman should not alone pursue personal interests but also advance the welfare of society as a whole (Sun et al., 2024).

The research indicates an imbalanced association between external and internal CSR performance in firms led by narcissistic CEOs. The justifications for this seem credible. The significant need for acknowledgment and media visibility is once again essential. External CSR performance, such as excellent communication and media presence, garners more attention than internal CSR projects. Narcissists typically prioritize their own demands, dismiss the views of others, and prefer to operate alone. Internal company initiatives for social and environmental performance are less disseminated and mainly target stakeholders. The most recognized internal stakeholder group is likely the workers. A successful CSR plan for this interest group would include equitable rights in all aspects, a diverse workplace, and several other advantages for workers. Narcissists are egocentric and assign less importance to the welfare of others. Consequently, they will use business resources to enhance their external influence and reputation. Al-Shammari et al. (2019) demonstrate a definitive positive correlation between CEOs' narcissistic tendencies and external CSR initiatives, while concurrently indicating a negative correlation between narcissism and internal company CSR performance. This association does not attain statistical significance, indicating that no definitive conclusion can be drawn on this matter Pizzetti et al., 2021. The upper echelon hypothesis, together with the results of Chin et al. (2013) and insights on narcissism, indicates that CEO personality qualities are predictors of greenwashing occurrences.

1.2.3 Board Gender Diversity

Gender equality, inclusion, and cultural and sexual diversity are among the numerous concepts that are essential in contemporary society with regard to diversity. These values have, for justifiable reasons, become deeply ingrained in Western culture. The literature indicates that equitable opportunities within the labor market not only enhance the welfare of employees but also have the potential to augment organizational performance. This indicates that the matter of diversity has garnered the attention of all stakeholders, including legislative bodies. Concerning legislative matters, it is important to highlight that numerous countries are currently deliberating the implementation of legally mandated quotas for women. Nevertheless, this has been implemented in a limited number of countries. Nonetheless, nations such as Norway are exemplifying leadership in this domain by instituting a mandate requiring a minimum of 40% representation of women on corporate boards of directors as early as 2003. The cautious stance adopted by the majority of nations necessitates that social considerations, on one hand, and economic advantages, on the other, serve as the primary catalysts for transformation within corporations and board committees (Peng & Xie, 2024; Yu et al., 2020; Zhao et al., 2024).

Upon examining both the upper stratum theory and the gender socialization theory, one may ponder the necessity of implementing a mandatory quota to guarantee diversity. The upper echelon theory posits that a diverse workforce can contribute a broad spectrum of knowledge, values, experiences, and cognitive approaches to the decision-making process. These perspectives, especially concerning the psychological distinctions between men and women, are similarly manifested in the theory of gender socialization. Advocates of this theory contend that women exhibit heightened sensitivity to manipulation, resulting in a pronounced sense of justice that compels them to confront inconsistencies and controversies directly, rather than disregarding them. Furthermore, the theory posits that individuals of the female gender exhibit a heightened sense of responsibility regarding social and environmental considerations within the organization.

Conversely, critics rationalize these perspectives as products of social construction. For instance, it is contended that children are socialized into the values, norms, and behaviors associated with their respective genders from the moment of birth. Irrespective of whether the psychological distinctions between men and women have been influenced by societal norms and expectations or are innate from birth, the literature provides evidence that women exhibit a greater propensity for ethical and morally responsible behavior. Consequently, they are more inclined to consider the interests of stakeholders. In light of these theories, the presence of female board members enhances the composition of the board, thereby rendering quotas unnecessary (Teti et al., 2024; Zhou et al., 2024).

Both the upper echelon theory and the gender socialization theory offer compelling evidence suggesting that female board members are potentially more inclined to advocate for the rights of stakeholders and exhibit a greater propensity for risk aversion and ethical conduct compared to their male counterparts. Owing to these attributes, women may possess a greater capacity to execute the fundamental responsibilities of a board, specifically providing guidance and oversight, in a manner that is more ethically sound. This aligns with the principles of the agency theory. In accordance with agency theory, board members are appointed by shareholders to mitigate opportunistic behavior that may contravene the economic interests of the shareholders by means of oversight and surveillance. Should a company choose to engage in greenwashing, it may incur adverse repercussions, including reputational harm, a decline in legitimacy, or financial liabilities, particularly if such practices are brought to light. When one examines the character characteristics typically associated with femininity, female board members may perceive it as excessively hazardous to amplify their corporate social responsibility initiatives in the media. The potential for publications to undergo scrutiny, along with the accompanying legal disputes or regulatory oversight, is ever-present. This propensity for risk aversion, coupled with the capacity to scrutinize potential instances of fraud and to reveal negligent managerial conduct, indicates that women may execute the monitoring function with greater efficacy than their male counterparts.

In light of these theories, it can be posited that organizations with a greater representation of women on their boards experience a reduction in information asymmetry between shareholders and the company, thereby mitigating the potential for greenwashing. Gull et al. (2023) investigate this phenomenon and conclude that women exhibit greater ethical and moral behavior, demonstrating a heightened consideration for the interests of stakeholders. This indicates that they do not disregard opportunistic behavior, thereby fostering transparency in accordance with the principles of the agency theory. The authors thus discover that board gender diversity plays an essential role in both internal and external CSR performance. The presence of women on the board of directors is inversely correlated with the phenomenon of CSR decoupling. Gull et al. further distinguish between female board members and female executive directors (Gull et al., 2023). The authors contend that independent directors experience diminished pressure from management in comparison to their counterparts who assume executive roles, and consequently, they are more inclined to implement measures that uphold corporate governance. In the research conducted by Gull et al. (2023), the assertion is substantiated that only independent female board members effectively mitigate the phenomenon of CSR decoupling. Furthermore, the study indicates that the inclusion of a single female board member amidst an otherwise male-dominated board results in only negligible alterations in the decision-making processes due to the prevailing influence of the minority. The research indicates that a minimum of two or three female board members, contingent upon the total number of board members, is necessary to effectuate a significant influence on corporate social responsibility (CSR) policy.

Gull et al. (2023) exclusively investigate the impact of board gender diversity on decoupling, whereas Eliwa et al. (2021) make a distinction between greenwashing and brownwashing. In accordance with the upper stratum and board gender socialization theory, the authors ascertain that boards predominantly composed of women are less inclined to partake in greenwashing, attributable to the distinct characteristics previously delineated. Concerning the phenomenon of brownwashing, Eliwa et al. (2021) identified no significant effects; consequently, this determinant will not be further addressed in the section pertaining to brownwashing within this book (Eliwa et al., 2021). When examining gender socialization theory through the lens of the notion that the distinctions between men and women are socially constructed, it becomes evident that religion exerts a considerable influence. Particularly in nations where religion is regarded as a quasi-legal framework, conventional gender roles continue to dominate. According to these perspectives, the male gender predominates in decision-making processes, potentially diminishing the influence of women's positive character characteristics in boardroom decisions. Eliwa et al. (2021) ascertain that the adverse impact of female-dominated management bodies on greenwashing is mitigated by the influence of religion. This suggests that a significant representation of women on the board is effective in mitigating greenwashing solely in companies situated in countries that uphold secularism (Eliwa et al., 2021; Gull et al., 2023).

The results presented by Xia et al. (2023) align with the previously examined research, indicating that male CEOs exhibit greater caution regarding penalty payments in financial reporting, which consequently leads to the concealment of information (Xia et al., 2023). Xia et al. (2023) contend that women exhibit a greater capacity for empathy, rendering them more attuned to social and environmental initiatives. Consequently, female board members exhibit a heightened interest in the implementation of sustainable practices within the organization, as well as in the promotion of transparent reporting. Consistent with the studies previously examined, Chen and Dagestani (2023) also identified a substantially negative correlation between the presence of female board members and the greenwashing attributes of companies (Chen & Dagestani, 2023). On the one hand, the attributes ascribed to the upper echelon theory, including stakeholder orientation and a long-term perspective, are essential for this purpose. Conversely, gender discrimination in China may subject female board members to heightened hostility, which they endeavor to mitigate by embracing ethical business practices. The research conducted by Chen and Dagestani (2023) not only supported prior studies but also demonstrated that boards comprised of younger and more highly educated members are associated with a diminished likelihood of engaging in greenwashing. In conclusion, the diversity of a board serves as a constraining factor in the practice of greenwashing (Fella & Bausa, 2024; Li, 2024; Pizzetti et al., 2021; Testa et al. 2018).

1.2.4 CEO Power

The term "entrenched manager" denotes a situation in which the managing director or board of a company is significantly entrenched within the organization as a result of specific circumstances. Examples of this phenomenon include elevated levels of specialized knowledge, as well as networks and relationships deemed essential for the organization, often referred to as "managerial entrenchment." The aforementioned implies that the manager is deeply entrenched within the organization, either due to extensive tenure or their specific role, which consequently renders them challenging to replace. Ultimately, the process of anchoring results in this individual exerting significant control and authority, thereby enabling them to shape the company's direction in accordance with their own perspectives (Kim & Lyon, 2015; Koch & Denner, 2025). Consequently, this may lead to the establishment of objectives that optimize their personal advantage and fortify their standing within the organization. Even if these actions do not align with the best interests of the company. The "entrenched manager" endeavors to obscure management decisions that are deemed inefficient from the company's perspective, along with the personal benefits that accrue to the manager therefrom. Such actions result in a diminished standard of corporate governance, thereby engendering a conflict as delineated by the principal–agent theory. This conflict is evident in the proprietors assuming the role of the principal, thereby delegating authority to the board or managing director as agents tasked with administering the organization in a manner that is both profitable and aligned with ethical and moral standards (Li et al. 2024; Zhang, 2022). Owing to the principal's insufficient oversight, the emissary capitalizes on his expertise and engages in opportunistic behavior.

In instances where "entrenched managers" exhibit behavior that is contrary to ethical and moral standards, there is an effort to adopt shareholder-friendly practices as a means of obscuring the true nature of their actions. A widely recognized practice in this context is referred to as "earnings management." In this context, the accounting regulations are examined to see if their legal boundaries are surpassed in pursuit of objectives or to fulfill the expectations of shareholders. When accounting regulations are modified to benefit the company, this practice is termed "accrual-based earnings management." The parameters established within the accounting standards are utilized, for instance, to recognize sales or expenses either prematurely or belatedly, thereby modifying business metrics in accordance with predetermined objectives and anticipations. Conversely, when tangible operational decisions are executed, such as deferring investments with the objective of maintaining elevated profit levels, this practice is termed "real earnings management" (Paganou et al., 2024; Ragazou et al., 2024; Zournatzidou et al. 2024a, 2024, 2024, 2024b).

García-Sánchez et al. (2022) discovered a substantially positive relationship between "accrual-based earnings management" and CSR decoupling in their study (García-Sánchez et al., 2022). This suggests that incumbent managers may exert influence over accounting standards to portray their social and environmental performance in a manner that is more favorable than the reality. The pivotal consideration in this context may be the considerable financial implications associated with Corporate Social Responsibility (CSR) initiatives. Considering the financial demands associated with Corporate Social Responsibility (CSR) initiatives, managers encounter a significant challenge. On the one hand, managers seek to leverage the favorable perceptions that shareholders hold regarding corporate social responsibility (CSR) initiatives in order to obscure the true nature of business practices. Conversely, investments associated with Corporate Social Responsibility (CSR) tend to be expensive and necessitate an extended repayment period. To sustain their status as "entrenched managers," it is plausible that managers engage in corporate social responsibility (CSR) activities that are predominantly symbolic in nature, thereby permitting a dissonance between the company's rhetoric and its actual practices. In their research, García-Sánchez et al. fail to distinguish between greenwashing and brownwashing (García-Sánchez et al., 2022).

Considering their lines of reasoning, it is evident that they suggest a positive correlation between "entrenched managers" and the prevalence of greenwashing. This is manifested in the authors' statements, whereby CSR decoupling is used to conceal unscrupulous business practices and to erect a "green" facade in front of stakeholders. Furthermore, the authors contend that Corporate Social Responsibility (CSR) is employed to create an overly favorable representation, while the company's senior management exerts influence over accounting regulations to perpetuate its autonomous decision-making. Entrenched managers, therefore, employ greenwashing to establish an information asymmetry in accordance with the principal–agent theory, thereby facilitating the execution of opportunistic actions (Kim & Lyon, 2015; Testa et al. 2018; Tsukada et al., 2023).

Nevertheless, "entrenched managers" represent merely one of the numerous subcategories that are associated with elevated CEO authority. In addition to the previously referenced principal–agent theory, the managerial power theory also elucidates this issue. This suggests that boards of directors, managing directors, or managers may leverage their authority to promote their personal interests rather than prioritizing those of the stakeholders. This also encompasses the notion that, akin to "entrenched managers," CEOs possessing significant authority and competence may similarly engage in opportunistic behavior concerning corporate social responsibility (CSR). Gull et al. (2023) investigate this thesis and identify a positive correlation between CEOs exhibiting high levels of power competence and the practice of greenwashing (Booth et al., 2024; Ghafoor & Gull, 2024). The authors identify a negative correlation concerning the phenomenon of brownwashing. The authors only discover the positive relationship between high exercise of power and greenwashing in companies with low levels of corporate governance. Furthermore, the research indicates that the sector in which the company functions is also of significant importance. Gull et al. (2023) present evidence indicating that the Chief

Executive Officers (CEOs) of firms within environmentally detrimental industries are subjected to heightened pressure from stakeholders. Consequently, these executives may resort to greenwashing practices in an attempt to project a seemingly reduced environmental impact. In contrast to the findings of García-Sánchez et al. (2022), the results presented herein pertain exclusively to environmental factors rather than social considerations (Huybrechts et al., 2024). Furthermore, Sauerwald and Su (2019) identify that the influence of Chief Executive Officers (CEOs), specifically those exhibiting self-assured and conceited personality traits, is positively correlated with opportunistic behaviors, particularly in the context of symbolic Corporate Social Responsibility (CSR) initiatives. Corporate governance mechanisms, including diversified ownership and the inclusion of additional board members possessing expertise in corporate social responsibility (CSR), may undermine this favorable relationship. In light of this empirical evidence, the influence of CEO power is unequivocally a factor in the occurrence of greenwashing. Furthermore, a potential correlation with brownwashing cannot be dismissed, which is why it is addressed in the corresponding section on brownwashing (García-Lillo et al., 2023; Huybrechts et al., 2024).

1.2.5 Global Reporting Initiative as a Uniform Set of Rules

The legitimacy theory posits that a company is deemed legitimate if it conforms to the set of values established by society. Considering the growing awareness among the populace regarding environmental and social policy matters, it is imperative for corporations to take into account not only economic factors but also ecological considerations in their operations, as posited by the legitimacy theory. Consequently, the execution of Corporate Social Responsibility (CSR) initiatives and their subsequent reporting is imperative for organizations to be perceived as legitimate within society (Kim & Lyon, 2015; Testa et al. 2018; Tsukada et al., 2023). It is important to acknowledge that most of the reporting is conducted on a voluntary basis and lacks a standardized set of regulations. This implies that organizations possess considerable latitude in their reporting practices, which may consequently result in the presentation of a potentially skewed representation of reality to their constituents. This problem has led to CSR reporting, which was once regarded as positive, now being viewed critically by stakeholders.

The Global Reporting Initiative (GRI) aims to address this issue. This framework, established by an independent international organization, offers corporations a set of guidelines for structuring their reporting on environmental, social, and governance issues, thereby ensuring a high degree of transparency and quality in their disclosures. The objective of the GRI is to limit the scope of corporations in their sustainability reporting mentioned above by means of an imposed reporting framework. This initiative is expected to enhance the consistency and comprehensibility of reporting across various organizations. While the literature highlights the favorable aspects on one hand, other studies express criticism toward the Global Reporting Initiative. It is contended that adherence to the mandated reporting standards primarily functions as a means of impression management, ultimately aimed at securing legitimacy (Bais et al., 2024; van Oorschot et al., 2024). Behnam and MacLean (2011) find evidence in their study of the symbolic application of the GRI. The findings indicate that organizations benefit from seemingly compliant reporting; however, in reality, only a subset of the services mentioned in the report have been effectively implemented (Vázquez-Ingelmo et al., 2024). From this perspective, organizations may encounter opportunities for opportunistic behavior in their reporting practices, even when such reporting ostensibly adheres to GRI standards (Liu et al., 2024; Nyilasy et al., 2014).

Conversely, the research conducted by Garcia-Sanchez et al. (2021) identified an inverse correlation between the utilization of the Global Reporting Initiative and the phenomenon of CSR decoupling (Grijalvo & García-Wang, 2023; Vázquez-Ingelmo et al., 2024). Consequently, the findings align with those of Sauerwald and Su (2019), who similarly identified diminished CSR decoupling as a consequence of adherence to GRI standards in their research. These studies indicate that the Global Reporting Initiative transcends mere symbolic guidance aimed at enhancing legitimacy. Conversely, adherence to the GRI standards in reporting enhances the quality of disclosures, fosters transparency, and subsequently mitigates the principal–agent conflict. Garcia-Sanchez et al. (2021) and Sauerwald and Su (2019) use the term CSR decoupling in their research without explicitly mentioning the phenomena of greenwashing or brownwashing (Grijalvo & García-Wang, 2023; Vázquez-Ingelmo et al., 2024). Nonetheless, the linguistic intricacies in the study results reveal, in light of the legitimacy theory, that the Global Reporting Initiative has a constraining influence, especially with greenwashing. Nonetheless, it can be deduced from the presented arguments that the implementation of regulated uniform reporting standards may also mitigate discrepancies between the verbal commitments and actual practices of companies. To prevent redundant processing, this determinant is not referenced again in the brownwashing part of this book.

Although the two studies previously mentioned do not explicitly address the phenomenon of greenwashing, as noted, implications can be inferred from their arguments. In contrast, the research conducted by Ruiz-Blanco et al. (2022) directly investigates the implications of the Global Reporting Initiative (GRI) reporting standard on greenwashing (Blanco-Davis & Zhou, 2016; Miotto et al., 2020). The authors contend that the issuance of sustainability reports establishes a communication conduit between firms and their stakeholders. Frameworks like the Global Reporting Initiative assist firms in aligning their reporting with stakeholder expectations and obligations, therefore mitigating the principal–agent conflict. Ruiz-Blanco et al. (2022) identify an inverse correlation between the GRI reporting standard and greenwashing.

1.2.6 Review of CSR Reports

The preceding paragraph has empirically demonstrated the significance of the Global Reporting Initiative in the context of CSR decoupling. The subsequent section will examine the auditing of Corporate Social Responsibility (CSR) reports conducted by independent third parties and their ramifications concerning greenwashing. In a manner akin to the obligatory auditing of annual reports by large corporations, partnerships, and financial and insurance institutions, stakeholders frequently seek the validation of Corporate Social Responsibility (CSR) reports through independent auditors. Proponents assert that external auditing serves to unveil opportunistic behavior, thereby diminishing information asymmetry among stakeholders and ultimately enhancing trust in reporting. Conversely, critics categorize the audit as a manifestation of greenwashing employed by corporations seeking legitimacy to sway stakeholders. Indeed, akin to Corporate Social Responsibility (CSR) reporting, the auditing of such reports is conducted within a predominantly unregulated framework that lacks clearly delineated processes (Li, 2024; Pizzetti et al., 2021). This raises concerns regarding the genuine efficacy and reliability of the verification process, as well as the performance metrics delineated in the sustainability report (Martínez-Ferrero et al., 2015). In contrast to the conclusions drawn by Sauerwald and Su (2019) and Ruiz-Blanco et al. (2022), both of which identify a constraining function associated with the verification process, Garcia-Sanchez et al. (2021) present evidence suggesting that the verification of Corporate Social Responsibility (CSR) reports is characterized by greenwashing tendencies (Ruiz-Blanco et al., 2022). Nevertheless, the findings do not surpass the established significance threshold. Nonetheless, the findings suggest that the concept of reputation, rather than a diminishment of the principal–agent conflict, frequently serves as a motivating factor for corporate verification. On the one hand, the findings of Garcia-Sanchez et al. (2021) elucidate that the pursuit of legitimacy may serve as a significant impetus for verification, thereby suggesting that the concept of greenwashing cannot be dismissed (García-Sánchez et al., 2022; Martínez-Ferrero et al., 2015). Conversely, the research conducted by Sauerwald and Su (2019), along with that of Ruiz-Blanco et al. (2022), suggests that an examination of reports by seasoned and specialized auditing entities may reveal discrepancies pertaining to greenwashing (Ruiz-Blanco et al., 2022).

1.2.7 Linguistic Design of the CSR Reports

In light of the growing volume of corporate publications addressing social and environmental issues, the implementation of a framework such as the Global Reporting Initiative (GRI) is of paramount importance. Nevertheless, even when a company adheres to the GRI guidelines, the linguistic construction of the reports may introduce a bias among stakeholders (Huang et al., 2024; Li, 2024; Liu et al., 2024; Pizzetti et al., 2021). Wei et al (2023) presents evidence indicating that reports characterized by positive information are articulated by companies utilizing accessible vocabulary. Conversely, reports that convey adverse implications for stakeholders are articulated in intricate linguistic patterns. The objective of the employed rhetoric is to obscure the true information, thereby mitigating the likelihood of a negative response.

In a more specific context pertaining to the study, Wang et al. (2018) investigated the relationship between the linguistic composition of Corporate Social Responsibility (CSR) reports and the corresponding CSR performance. The study indicates that companies exhibiting strong corporate social responsibility (CSR) performance utilize a more accessible level of language in their reports. Conversely, organizations that fail to undertake meaningful actions often employ language designed to divert stakeholders' attention from the underlying issues, utilizing convoluted sentence structures and esoteric terminology. Furthermore, Wang et al. (2018) discovered that a sophisticated linguistic style is employed solely for the symbolic social advantages of the organization (Koch & Denner, 2025). This may be attributable to the comparatively stringent regulations governing environmental performance, as delineated by various ISO standards. Social initiatives, for instance, are likewise governed by the ISO standards. It is important to acknowledge, however, that these ISO certifications are less prevalent in the market compared to those pertaining to environmental performance, which may render stakeholders more susceptible to misinformation. Through the utilization of intricate linguistic constructs, corporations endeavor to obfuscate their stakeholders' understanding, thereby complicating the distinction between the company's substantive and symbolic social initiatives. The company is therefore selectively disseminating information, emphasizing favorable elements (Hussain et al., 2023; Wang et al., 2018). The terminology employed in Corporate Social Responsibility (CSR) reports, especially concerning social advantages, is consequently a critical factor in the assessment of greenwashing.

1.2.8 CSR-Komitee

In addition to numerous other responsibilities, the board of directors is tasked with overseeing and providing counsel to the company. Consequently, the board members assume a pivotal function in the governance of the corporation. The objective of this function is to represent the interests of the shareholders in the best possible manner, thereby limiting the information asymmetry between the company and the interest groups to a minimum. It is important to acknowledge that board members, akin to all individuals, are influenced by specific character characteristics. The diverse psychological distinctions have previously been addressed within the framework of the upper echelon theory, as well as in earlier sections, including Sect. 1.2.2. These characteristics indicate that the company's actions do not consistently align with the expectations of its stakeholders. There frequently exists a significant disparity, particularly concerning corporate social responsibility (CSR) initiatives. In the realm of greenwashing, corporations present a façade of social responsibility and sustainable performance to their shareholders. Nevertheless, owing to the principal–agent dilemma, shareholders frequently find themselves unable to perceive the underlying realities behind this façade. Information asymmetry implies that shareholders are unable to discern the symbolic significance of these services (Cahan et al., 2016; Faieq & Cek, 2024). Consequently, this allows management to sustain its opportunistic conduct without impediment. In light of this consideration, it is imperative to establish independent boards of directors within the company that are especially dedicated to addressing social and environmental issues both internally and externally. Yu et al. (2020) similarly express this perspective in their research. In their research, the authors conclude that board members who possess a significant degree of independence from the company, in conjunction with an increased overall number of board members, can enhance the effectiveness of the monitoring function. Yu et al. (2020) further contend that an increase in the number of board members may, in certain contexts, pose challenges. In addition to the existence of free riders, coordination challenges frequently emerge among board members, which can consequently result in protracted decision-making processes (Amin et al., 2020; Nasrullah, 2011). To potentially mitigate this issue, the establishment of a distinct and autonomous Corporate Social Responsibility (CSR) committee may be considered.

In precise terms, the concept of a Corporate Social Responsibility (CSR) committee entails the establishment of a dedicated focus group that specifically addresses social and sustainability issues, encompassing their development, implementation, and oversight. To effectively address these points, as articulated by Godos-Diez et al. (2018), the focus group should comprise experts—individuals who possess substantial knowledge and expertise in the relevant field. To ensure the provision of adequate expertise, it is reasonable to assert that the size of the committee is also of paramount importance (Faieq & Cek, 2024). On the one hand, an increased membership is correlated with enhanced value in terms of experience, improved diversification of training and skills, as well as greater capacity for oversight. However, the challenges previously identified by Yu et al. (2023) also pose significant obstacles in this context. Gull et al. (2023) investigate this dual-faceted phenomenon and conclude that a substantial Corporate Social Responsibility (CSR) committee is more likely to possess the capacity to mitigate greenwashing, attributable to its inherent diversity.

It is also debatable whether the composition of the CSR committee should include individuals from within the organization or external stakeholders. On the one hand, Godos-Diez et al. (2018) argue that a CSR committee consisting of internal stakeholders may make it simpler to establish the strategy intended by the committee in the organizational processes. This is attributable to the fact that internal stakeholders, including employees, possess a comprehensive understanding of the company's operational processes, organizational structures, and corporate culture (Liu et al., 2024; Liu et al., 2025; Xu et al., 2024). Consequently, they are better positioned to align the Corporate Social Responsibility (CSR) strategy with the company's objectives than external interest groups. Similarly, the participation of employees in the committee fosters their commitment to and approval of the CSR-specific initiatives, thereby enhancing the likelihood of successful implementation and sustained adherence within the organization. Conversely, Godos-Diez et al. (2018) contend that the members of the Corporate Social Responsibility (CSR) committee ought to maintain independence from the company. This implies that participants must not possess any personal or financial dependence on the company;

otherwise, their decisions may lack objectivity and impartiality. In light of this, and in accordance with the research conducted by Gull et al. (2023), it has been observed that members who are independent of the company exhibit a heightened concern for corporate social responsibility (CSR) issues. Consequently, it is advisable to convene a focus group comprising external stakeholders (Henao-Rodríguez et al., 2024; Oguntona, 2024; Peng & Xie, 2024; Sun et al., 2024).

If, on the other hand, we now consider a CSR committee consisting of internal stakeholders, the length of service is essential. From this point of view, it can be argued that a long period of employment helps members to gain company-specific experience, which can be used to successfully combat greenwashing. On the other hand, a long employment relationship may bring the member into conflict with the independence described above. From this perspective, it can be argued that a long connection with the company could lead to limited monitoring, and thus, according to the previous argument, external stakeholders should be preferred (Chen et al., 2024; Zournatzidou, Ragazou, Sklavos, et al., 2024a).

In Sect. 1.2.3, the impact of gender diversity within the board structure on the phenomenon of greenwashing was analyzed. Eliwa et al. (2021) discovered evidence indicating that a considerable percentage of women in management positions serve to mitigate greenwashing practices (El Koshiry et al., 2023; Eliwa et al., 2021). In this context, a comparable relationship may also be envisaged concerning the composition of the Corporate Social Responsibility (CSR) committee. Concerning the focus group, Gull et al. (2023) discovered no statistically significant correlation between the biological sex of the participants and the successes attained.

The findings of Gull et al. (2023) concerning the Corporate Social Responsibility (CSR) committee can be regarded as seminal contributions to this subject. In accordance with the principal–agent theory, the authors assert that the formation of a Corporate Social Responsibility (CSR) board committee constitutes an effective mechanism for corporate governance. This implies that a focus group specifically geared toward social and sustainable aspects helps to improve both the implementation and monitoring of CSR-specific aspects (Booth et al., 2024; Ghafoor & Gull, 2024). In the introduction to their study, Gull et al. (2023) make a distinction between greenwashing and brownwashing; however, throughout the course of their research, particularly in their findings, they fail to maintain this differentiation between the positive and negative manifestations of decoupling. This indicates that the findings are pertinent to both greenwashing and brownwashing.

Although the findings of Gull et al. (2023) demonstrate statistical significance concerning the efficacy of a Corporate Social Responsibility (CSR) board committee and align with the principal–agent theory, the existing literature remains divided on this issue. Haque and Ntim (2020) present a divergent perspective on the Corporate Social Responsibility (CSR) committee. The study, which investigates the compensation of executives based on ecological rather than economic criteria, posits that a Corporate Social Responsibility (CSR) committee functions primarily as a symbolic gesture (Karagiannis et al., 2022; Liu et al., 2020; Ma & Ahmad, 2024; Nasrullah, 2011; Petrenko et al., 2016). The authors contend that companies establish a Corporate Social Responsibility (CSR) committee to highlight their social and environmental performance in a manner that is effective for media engagement. Aside from the enhanced discourse that results from this perspective, there is minimal alteration in the company's operational practices. This further underscores that the establishment of a Corporate Social Responsibility (CSR) committee is a significant advancement; however, the composition and attributes of the focus group are equally critical to its efficacy. A Corporate Social Responsibility (CSR) committee that adheres to the aforementioned characteristics can be regarded as a constraining factor in mitigating both greenwashing and brownwashing practices (Nepal et al., 2025; Nyilasy et al., 2014; Yao et al., 2024).

1.2.9 Normative Pressure from Stakeholders

Upon examining a company's stakeholders, one can categorize them into two primary categories. On one hand, external interest groups, including non-governmental organizations (NGOs), consumers, financial intermediaries, and industry associations, exert a degree of influence on corporations. Conversely, internal constituents, including managers and employees, play a pivotal role in shaping the norms and values of an organization. In accordance with institutional theory, these two interest groups predominantly influence the strategic direction and corporate social responsibility (CSR) policies of the organization. Due to the pressure to conform that stakeholders impose on the company, organizations experience a genuine pressure to satisfy the demands of their interest groups. In this manner, the company implements measures that are regarded by society as suitable in order to fulfill its expectations. This compulsion to conform is also characterized as self-regulation or normative stakeholder pressure (Kiran et al., 2024; Tao, 2025).

Institutional theory posits that organizations are influenced by the normative pressures exerted by their constituents, which consequently leads them to adopt specific actions. Within the framework of greenwashing, it may be posited that organizations may superficially fabricate certain corporate social responsibility (CSR) initiatives in response to normative pressures, thereby aligning their actions with prevailing societal expectations (El Koshiry et al., 2023). Nevertheless, a significant implementation of social and environmental performance is unlikely to be achieved. In the context of internal normative pressure, the norms, values, and attitudes of both employees and management hold significant relevance. Should these stakeholder groups express interest in the corporate social responsibility (CSR) issue, it is evident that socially and environmentally sustainable strategies will be effectively integrated into the daily operations of the company. It is also plausible that the implemented actions are communicated with transparency to subsequently obtain any potential advantages. Hyatt and Berente (2017) investigate this thesis and identify a positive correlation in both instances. The research indicates that internal normative pressure is more likely to result in the execution of both substantive and symbolic actions. This may be attributable to employees' awareness of the company's genuine stance on these matters. The primary objective of symbolic corporate social responsibility (CSR) initiatives is frequently to attain legitimacy. Should management opt to undertake solely symbolic actions, employees are likely to exhibit minimal acknowledgment and legitimacy. This indicates that internal pressure predominantly facilitates the effective execution of measures specific to Corporate Social Responsibility (CSR), which are subsequently communicated appropriately in the event of success. Consequently, internal normative pressure may be categorized as a factor that constrains greenwashing practices. Consequently, it is imperative to enhance awareness among internal stakeholders by implementing training courses or seminars that specifically address this issue, as this is crucial for the establishment of robust corporate social responsibility (CSR) practices.

Although the norms, values, and attitudes of both managers and employees hold significant importance within the organization, there exists a multitude of external stakeholders with diverse and varying expectations of the company. No single company is capable of fulfilling all of these requirements. In accordance with institutional theory and the normative stakeholder pressures previously delineated, the organization may engage in self-regulation concerning corporate social responsibility (CSR) to fulfill the expectations of the broader public. This indicates that certain methodologies are, in fact, executed in practice. A significant proportion, however, is conveyed to the public solely in a symbolic manner. In contrast to internal stakeholders, external stakeholders possess limited access to the underlying realities that may lie beyond the company's public relations façade. The resultant agency conflict permits the company to fulfill only a portion of the requisite obligations while still maintaining a socially and environmentally responsible image in the eyes of external stakeholders. Hyatt and Berente (2017) present findings in their research indicating that external normative pressure results in both symbolic actions and, in certain instances, significant measures. Nevertheless, the research elucidates that the predominant aspect of this phenomenon is of a symbolic nature (Petrenko et al., 2016). Consequently, the findings are consistent with those of Testa et al. (2018). The study also revealed that external stakeholder pressure frequently results in "corporate greening," which can be interpreted as greenwashing. Only influential shareholders, financial institutions, or other intermediaries in the financial sector exert a positive impact on the effective implementation of sustainable practices. This phenomenon can be attributed to the enhanced control mechanisms that provide these stakeholder groups with a more profound understanding of the respective company, thereby facilitating the identification of symbolic actions.

In this context, it can be posited that financial considerations are often the primary impetus for engaging in greenwashing practices. Both the implementation of actions ("walk") and the accompanying discourse ("talk") pertaining to Corporate Social Responsibility (CSR) require substantial financial resources. Profit-driven enterprises must carefully evaluate whether they are implementing cost-intensive strategies or engaging in opportunistic investments in symbolic initiatives. The principal–agent conflict, along with the resultant information asymmetry, allows organizations to align their actions with the expectations of stakeholders in accordance with institutional theory. In this context, the company has the capacity to construct a deceptive façade, especially in relation to external stakeholders. The research conducted by Hyatt and Berente (2017), along with that of Testa et al. (2018), demonstrates that both internal stakeholders and specific external stakeholders possessing a regulatory function are able to discern this façade. Organizations recognize the potential legitimacy risks associated with this matter and may undertake significant corporate social responsibility initiatives in response to both internal dynamics and particular external pressures.

1.3 Consequences of Greenwashing

Given the numerous factors that either facilitate or hinder the prevalence of greenwashing, it is pertinent to examine the implications that greenwashing may have on the financial performance of companies as well as their environmental impact. The primary ramifications are delineated in the following sections.

1.3.1 Impact on Company Performance

To evaluate the influence of greenwashing on corporate performance, Schons and Steinmeier (2016) distinguish between stakeholders who maintain a close relationship with the company and those stakeholder groups that adopt a more distant stance, as informed by stakeholder theory. The primary determinant for classification lies in the understanding that the pertinent interest group gains insight into the company as a result of the prevailing principal–agent conflict. If we now examine the previous stakeholders, employees, for instance, can be categorized within this group. As members of the organization, they possess a notably comprehensive understanding of the company's operations, enabling them to effectively differentiate between symbolic and substantive corporate social responsibility (CSR) initiatives. In contrast, external interest groups, including customers and investors, experience an information asymmetry that complicates their ability to distinguish between symbolic and substantive corporate social responsibility (CSR) initiatives.

Consequently, Schons and Steinmeier (2016) identified evidence in their research indicating that a greenwashing strategy employed by stakeholders who maintain a distant relationship with the company correlates with enhanced financial performance. In contrast, the authors observe a detrimental financial performance attributable to greenwashing among direct stakeholders. The authors do not provide a detailed explanation of the term "financial performance" within the context of their study (Wang & Sarkis, 2017). It is evident from the methodological section that the authors utilized the Asset4ESG database for their research, which offers comprehensive insights into the financial performance of companies by incorporating accounting-based metrics such as Return on Assets (ROA), shareholder loyalty as measured by Return on Invested Capital (ROIC), and customer loyalty, assessed through customer satisfaction (Demers et al., 2021). In contrast, Walker and Wan (2012) do not distinguish between the two interest categories as delineated by stakeholder theory. As previously articulated in Sect. 2.1.15, the primary emphasis of this research pertains to the ramifications of greenwashing and its consequent impact on corporate performance within environmentally detrimental industries. Walker and Wan (2012) assert that greenwashing erodes stakeholders' confidence in the corporate social responsibility (CSR) performance of organizations, consequently diminishing their financial performance over the long term. At this juncture, however, it should be mentioned that Walker and Wan (2012) also reduce the term "negative financial performance" to the level of companies' ROA.

While the two papers discussed previously do not investigate the concrete specifics of financial performance, Cao et al. (2022) address the implications that greenwashing has on the cost of debt and the collateral to be deposited. Similarly, Cao et al. (2022) identify in their research that firms operating within emerging markets engage in greenwashing as a strategy to enhance their capital-raising capabilities. Specifically, they uncover evidence indicating that lenders extend more favorable interest rates and require reduced collateral from companies listed on the "Best CSR List." Nonetheless, their research indicates that inclusion on the "Best CSR List" does not serve as an accurate indicator of genuinely exceptional corporate social responsibility initiatives within the organization. Conversely, the authors present evidence indicating that the companies featured on the list allocate substantially greater resources to advertising expenditures compared to those not included in the list. This indicates that the companies are endeavoring to project an environmentally friendly image through substantial advertising expenditures, with the intention of subsequently reaping the financial benefits associated with inclusion in the "Best CSR List." In alignment with the findings of Xia et al. (2023), this study similarly demonstrates that the fragile institutional framework characteristic of emerging markets facilitates the prevalence of greenwashing, thereby enabling firms to gain advantages in capital acquisition.

In contrast to the research conducted by Cao et al. (2022), the investigation by García-Sánchez et al. (2022) explores the ramifications of greenwashing on capital acquisition within the United States of America. As previously stated, the provision of high-quality Corporate Social Responsibility (CSR) information plays a crucial role in enhancing the accuracy of analyst estimates. This, in turn, mitigates information asymmetry and diminishes the propensity for companies to engage in greenwashing practices. Investors are thus positioned to make objective assessments regarding the economic and ecological risks associated with the company, enabling them to establish correspondingly reduced capital costs. Should companies engage in greenwashing, it will adversely impact the accuracy of analyst estimates and exacerbate the principal–agent conflict. The heightened prevalence of greenwashing may lead investors to exercise caution and impose elevated premiums when allocating capital. This assertion is in direct conflict with the findings presented by Cao et al. (2022). Research indicates that, in emerging countries, greenwashing yields favorable outcomes for corporate performance, in contrast to its effects in established industrial nations, owing to the prevailing institutional conditions.

1.3.2 Impact on Company Value

An examination of the research conducted by Chen and Dagestani (2023) reveals that their findings indicate a correlation between greenwashing and an enhancement in corporate value. When one contemplates the fundamental tenets of stakeholder theory within this context, such an outcome is scarcely unexpected. This posits that stakeholder engagement within the organization should not merely be regarded as an ethical obligation but also serves as a strategic approach to cultivating competitive advantages over rivals. Upon examining the phenomenon of greenwashing, one may observe that this conduct, at least initially, appears to facilitate the exchange of information between corporations and their stakeholders. In contrast to substantive practices that may be challenging for interest groups to comprehend, greenwashing strategically formulates messages in an intuitive manner, thereby capturing the attention of stakeholders. In this context, Chen and Dagestani (2023) contend that the quality of published Corporate Social Responsibility (CSR) reports enhances investors' confidence, consequently influencing the company's market capitalization. The Kaplan–Zingales (KZ) Index indicates that an elevated index value correlates with a significant degree of transparency and quality in the realm of Corporate Social Responsibility (CSR). The findings of the study indicate that companies engaging in greenwashing attain elevated scores on the KZ Index, as well as significant increases in their market capitalization. Consequently, the misleading conduct results in an elevated perception of publication quality, heightened attention, alleviation of financial constraints, and ultimately enhances the company's value in accordance with stakeholder theory. If greenwashing is practiced by a company that operates in environmentally sensitive industries or obtains an audit from one of the Big Four auditing firms,151 this leads to increased stakeholder trust in the company and thus also to increased company value.

Hawn and Ioannou (2016) also provide evidence supporting a positive correlation between greenwashing and corporate value. They observe that a company's value is significantly enhanced when it engages in external actions rather than relying predominantly on internal initiatives. According to the findings presented by Hawn and Ioannou (2016), it is posited that a discrepancy of 0.02 in external communication relative to internal actions correlates with a market value of the company that is 45% greater than the average market value of its peer group. In comparison to companies exhibiting a discrepancy of zero, the "greenwashing" company consequently attains a company value that is 18% greater. Nevertheless, the research conducted by Hawn and Ioannou (2016) attains statistical significance solely within sectors that are particularly responsive to social and environmental performance, specifically the extractive and mining industries. Nevertheless, both studies fail to consider that companies that participate in greenwashing are perpetually susceptible to the risk of exposure. The author Du (2015) explores this subject through the concept known as the "backfire effect." This phenomenon pertains to investors scrutinizing companies that purport to engage in environmentally sustainable practices while, in actuality, engaging in unscrupulous business conduct that contradicts their publicly projected image. Should greenwashing be detected, the favorable reputation of the company will be compromised.

Investors have reverted to their initial stance, viewing the company not only as detrimental to the environment but also as hypocritical and lacking in transparency. As a result, the company experiences a deterioration of its image and reputation, leading to a decline in its overall value due to the practice of greenwashing, which contradicts the findings of the studies previously examined. Du (2015) substantiates his argument by demonstrating a substantial negative correlation between revealed greenwashing and the company's cumulative anomalous returns (CAR). The negative cumulative abnormal return (CAR) indicates that stakeholder confidence has been undermined due to the inconsistency between the company's rhetoric and its actions, leading to a decline in the company's valuation. This indicates that the initial favorable impact on the company's valuation resulting from greenwashing is ultimately overshadowed by the revelation of opportunistic conduct and the ensuing "backfire effect" (Peng & Kong, 2024). Liu et al. (2023) present findings in their literature review indicating that greenwashing may adversely affect corporate value, even in the absence of detection. Their argument is predicated on the assertion that external corporate social responsibility (CSR) services, including comprehensive reporting, enhance external legitimacy, thereby mitigating the principal–agent conflict. Consequently, the share price experiences reduced fluctuations, thereby mitigating substantial losses for investors and minimizing volatility in the company's market capitalization. Nevertheless, the implementation of internal Corporate Social Responsibility (CSR) measures is essential for enhancing corporate value. These measures guarantee that the constituents of the company, as well as the environment, genuinely derive advantages from the proclaimed social and environmental initiatives. Liu et al. (2023) discovered that when the relationship is incongruent and the "walk" is inadequately developed, the advantages of external reporting remain unattainable due to a lack of stakeholder support and the absence of significant reputation-building initiatives. Consequently, it can be inferred from the two studies that greenwashing adversely affects corporate value.

1.3.3 Consumer Perception

In addition to traditional Corporate Social Responsibility (CSR) reports, a well-structured company website is pivotal in facilitating engagement with stakeholders, including the organization's clientele. As previously stated, the research conducted by Szabo and Webster (2021) demonstrates that the interactivity of homepages plays a crucial role in facilitating stakeholders' engagement with Corporate Social Responsibility (CSR) content. This enhanced engagement leads to a more profound comprehension of the presented material and, consequently, a reduction in skepticism regarding greenwashing. If stakeholders nevertheless discern a discrepancy between "walk and talk," the authors record negative impressions from visitors. In this context, the research conducted by Nerb and Spada (2001) demonstrates that stakeholders' wrath concerning environmental issues frequently results in a propensity to boycott the company. The research conducted by Ekman and Cordaro (2011) ought to be referenced as well. This indicates that the perception of specific cognitions, such as greenwashing, can elicit negative sentiments that are discernible through facial expressions. Szabo and Webster (2021) also showed that confrontation with greenwashing is negatively correlated with the satisfaction of the test subjects. Greenwashing resulted in strained facial expressions, characterized by furrowed brows, glistening eyes, and elevated corners of the mouth, all of which signify the fundamental emotion of wrath. According to the study, the perception of greenwashing does not alter the documented interactions of the test subjects, suggesting that there is no indication of a boycott against the company as a result of greenwashing. Nevertheless, it results in adverse facial expressions, thereby influencing the affective state of the participants in the study.

In addition to the findings of Szabo and Webster (2021), Ioannou et al. (2022) similarly identified a negative correlation between greenwashing and customer satisfaction. This phenomenon occurs because stakeholders categorize perceived greenwashing as a form of duplicity. The company's stance is hypocritical, as it advocates for a specific commitment through its Corporate Social Responsibility initiatives. Should stakeholders identify deficiencies manifested as greenwashing, not only are their economic interests jeopardized, but such revelations also engender a conflict that contravenes their ethical and moral expectations of the organization. If the organization is perceived as duplicitous, its value is considerably diminished from the standpoint of the consumer. If consumers perceive a diminished value in their relationship with the company, their level of satisfaction is consequently diminished. Greenwashing results in a negative evaluation and perception of the company among consumers, thereby diminishing their overall satisfaction. This matter warrants critical examination, as customer satisfaction serves as a pivotal factor influencing a company's financial performance. Companies can only impose exorbitant prices on consumers when customer satisfaction is elevated.

An examination of the literature review conducted by Chen and Chang (2013) reveals additional insights regarding the origins of the negative sentiments and diminished consumer satisfaction highlighted by Szabo and Webster (2021). The literature review indicates that the substantial volume of publications on Corporate Social Responsibility (CSR), in light of the stakeholders' constrained capacity to assimilate information, results in a significant excess of information. This, along with the considerable quantity and intricacy of the extant certificates examined in previous subsections, as referenced in the work of Montgomery et al. (2023a), renders it nearly impossible for consumers to distinguish between symbolic and substantive corporate social responsibility (CSR) initiatives. This indicates that consumers cultivate a sense of skepticism regarding Corporate Social Responsibility (CSR) communications, as they consistently perceive a degree of "green perceived risk." Chen and Chang (2013) interpret this as the risk that consumers, through their purchasing behaviors, inadvertently contribute to adverse social and environmental consequences that are contrary to their genuine intentions. In the event that consumers are unable to form a sound assessment regarding the social and environmental compatibility of a product, a detrimental trust relationship emerges, which the study designates as insufficient "green trust." The research conducted by Chen and Chang (2013) demonstrates that greenwashing amplifies perceived risk by generating confusion while concurrently eroding consumer trust in companies that are committed to corporate social responsibility (CSR).

The argument posited by Chen and Chang (2013) appears to be fundamentally sound, asserting that greenwashing, manifested through deceptive publications, engenders perplexity and, consequently, a lack of trust among consumers. Nevertheless, upon examining the research conducted by Aji and Sutikno (2015), one observes an absence of a negative correlation between "green consumer confusion" and "green trust." Furthermore, the authors present evidence in their study indicating that perplexity, skepticism, and perceived risk associated with greenwashing do not influence the subjects' consumption behavior, leading them to shift from ostensibly "green" products to "non-green" alternatives. Nonetheless, upon examining consumer skepticism toward environmentally focused publications, which is attributed to greenwashing, Aji and Sutikno (2015) also identify that such skepticism adversely affects consumer trust. The authors suggest that a deficiency of trust in purportedly "green" products may serve as a factor influencing stakeholders to contemplate alterations in their purchasing behavior. When evaluating the somewhat contradictory findings from the two studies, it is essential to acknowledge that Chen and Chang (2013) conducted their research in Taiwan, while Aji and Sutikno (2015) focused on Indonesia. Although Taiwan is recognized as a highly developed industrial nation, Indonesia distinguishes itself as an emerging country, characterized by its existing legal and economic framework as well as its approach to corporate social responsibility (CSR). Nonetheless, the notion of "green marketing," along with the phenomenon of greenwashing, is also evident in Indonesia. Consequently, the research conducted by Aji and Sutikno (2015) can be regarded as pioneering in this context.

In contrast to earlier studies, Torelli et al. (2020) examine four distinct manifestations of greenwashing and their ramifications for stakeholders' perceptions of this phenomenon. The authors characterize "corporate-level greenwashing" as deceptive corporate social responsibility initiatives that pertain to the organization as a whole. Conversely, "product-level greenwashing" pertains to the misleading promotion of specific products or divisions within a company. "Strategic-level greenwashing" is defined as the deceptive representation of a company's corporate social responsibility (CSR) strategic initiatives and commitments. Ultimately, the authors characterize "dark-level greenwashing" as corporate practices designed to obscure illicit activities associated with corporate social responsibility (CSR). An examination of the findings presented in the study conducted by Torelli et al. (2020) reveals that corporate-wide greenwashing is likely to foster the perception among stakeholders that the organization is engaging in environmentally responsible practices. Similar to strategically oriented greenwashing, this practice tends to engender a diminished degree of mistrust among stakeholders, thereby enabling companies to enhance their legitimacy and reputation through symbolic gestures. The study indicates that even in the event of environmental controversies, one can anticipate a diminished response from stakeholders. Conversely, the research indicates that "dark-level greenwashing" is associated with organized crime, resulting in diminished trust in corporate statements and fostering skepticism among stakeholders regarding corporate social responsibility (CSR). The study further documents an elevated degree of skepticism among companies that partake in product-specific greenwashing and concurrently function within environmentally sensitive sectors. Section 2.1.15 has already elucidated that enterprises operating within environmentally sensitive sectors are subjected to notably rigorous oversight. Greenwashing, which pertains solely to particular facets of the organization, stands in stark contrast to the overall conduct of the company. Based on this line of reasoning, organizations that partake in selective greenwashing and possess a detrimental reputation due to their industry may face the peril of their actions being exposed by stakeholders, resulting in appropriate sanctions. Torelli et al. (2020) elucidate, in the context of environmental controversies, that companies engaged in "dark-level greenwashing" or operating within environmentally sensitive sectors should anticipate a robust response from stakeholders upon the revelation of such practices.

In contrast to prior research, Torelli et al. (2020) demonstrate that the manner of expression is critical for the identification of greenwashing and the ensuing response. The aforementioned studies indicate that the phenomenon of greenwashing elicits inherently negative sentiments, skepticism, and perceived risk among consumers regarding corporate practices, thereby diminishing their overall satisfaction. These findings are further corroborated by the literature review conducted by Yang et al. (2020, p. 167), which concludes that greenwashing has detrimental effects on consumers.

1.3.4 Impact on Reputation

In their investigation of the impact of greenwashing on corporate reputation, Jong et al. (2019) delineate between "behavioral-claim greenwashing" and "motive greenwashing." The former encompasses those companies that acknowledge a divergence between their verbal commitments and their actual practices. Conversely, the latter refers to organizations that engage in significant sustainable initiatives while inherently advocating for a motivation that differs from the one conveyed externally. An illustrative example of this phenomenon is that the company does not engage in Corporate Social Responsibility (CSR) initiatives due to intrinsic motivation; rather, it undertakes such projects solely to fulfill stakeholder expectations or to adhere to regulatory requirements. In this context, Jong et al. (2019) assert that greenwashing invariably has detrimental effects on a company's reputation. Reputational harm ensues irrespective of whether the incongruity between symbolic and substantive actions impacts the entirety or merely specific components of the corporate social responsibility initiatives. Furthermore, the compromised reputation is not solely associated with the social and environmental performance itself. Jong et al. (2019) ascertain that a negative image adversely influences the perceived quality of both products and services, in addition to impacting financial performance. The authors substantiate this assertion by referencing the halo effect, as documented in the existing literature. This phenomenon pertains to the inclination of individuals to infer conclusions regarding other, genuinely independent attributes based on acquired knowledge. In this context, this implies that when stakeholders become aware of greenwashing, they are likely to scrutinize the aforementioned specifications with equal rigor, despite the absence of any direct correlation to the issue at hand. Ioannou et al. (2022) further demonstrate in their research that perceived hypocrisy, manifested as greenwashing, diminishes the trustworthiness and, consequently, the reputation of the entire organization. The authors contend that a company possessing a competitive advantage, such as the superior quality of its products, may serve as a safeguard, thereby shielding the organization from potential reputational damage. This is attributable to the fact that stakeholders inherently seek to engage with an organization that possesses a notably elevated degree of legitimacy as a result of its competencies. For instance, within this context, it is conceivable that a company may provide exceptionally innovative and high-quality products. A confrontation concerning the alleged hypocrisy may nonetheless encourage stakeholders to maintain their allegiance to the company, owing to the competitive advantages outlined. In contrast to the findings presented by Ioannou et al. (2022), one might contend that appealing competitive advantages draw a substantial number of stakeholders, thereby enhancing oversight and consequently revealing opportunistic behavior, which in turn may constrain reputation. Similarly, stakeholders may have heightened expectations that the company fails to fulfill. Consequently, the diminished legitimacy and reputation attained would result in reduced stakeholder satisfaction, as posited by Ioannou et al. (2022). Ioannou et al. (2022) ascertain in their research that specific competitive advantages function as a safeguard against the potential erosion of legitimacy resulting from greenwashing.

Upon reevaluating the findings of Jong et al. (2019) regarding "motive greenwashing," the authors conclude that the disclosure of such motives results in an incremental reputational loss solely in instances where the organization has not previously faced condemnation for "behavioral-claim greenwashing." Concerning the ramifications for corporate reputation, a disparity between rhetoric and action assumes a considerably more substantial role than the mere revelation that the implemented corporate social responsibility initiatives predominantly aim to adhere to specific standards.

1.3.5 Impact on Employees

Upon examining social identity theory, it posits that employees exhibit a stronger identification with an organization when the values of the company align closely with their personal values. The embodiment of this social and moral identity results in the employer being regarded as prestigious and appealing, consequently enhancing employees' self-esteem. In light of the widespread prominence of Corporate Social Responsibility (CSR) within contemporary society, it may be inferred that employees experience a heightened sense of affiliation with an organization that adheres to ethical practices, as such conduct is indicative of its social and moral standing. However, organizations are cognizant of the advantages linked to Corporate Social Responsibility (CSR) initiatives, including enhanced employee satisfaction, commitment, well-being, and numerous additional benefits. Although certain companies implement meaningful corporate social responsibility (CSR) initiatives, the advantages associated with such actions may entice some organizations to engage in merely symbolic gestures. Nevertheless, internal positioning provides employees with direct insight into the company's genuine practices, thereby diminishing information asymmetry and facilitating the distinction between symbolic and substantive actions. In their study, Robertson et al. (2023) investigated the phenomenon of greenwashing and its implications for employees. The authors assert that, in accordance with "social identity theory," greenwashing stands in stark opposition to the moral values upheld by employees. This dissonance results in the perception of the company as hypocritical and leads to a diminished sense of alignment among individuals with similar values. The research further indicates that the level of education serves as a critical determinant in this context. Robertson et al. (2023) ascertain that individuals who perceive greenwashing, particularly those possessing advanced educational qualifications, such as a master's or doctoral degree, are more inclined to depart from organizations as a result of such practices. This assertion is grounded in the observation that individuals with higher levels of education and acquired knowledge are generally more adept at distinguishing between symbolic and substantive actions. The prerequisite for this, consequently, is that employees have acquired the requisite knowledge through their instruction. Irrespective of the knowledge acquired, it can be posited that individuals possessing this level of education have assimilated a set of values that engenders a heightened degree of disidentification with organizations that engage in greenwashing practices. The research conducted by Robertson et al. (2023) demonstrates that greenwashing elicits adverse reactions from employees, consequently resulting in heightened turnover rates, particularly among individuals with higher levels of education.

In alignment with the research conducted by Robertson et al. (2023), Li et al. (2022) similarly ascertain that the perception of greenwashing adversely affects employee morale. In their study, Li et al. (2022) provide evidence indicating that greenwashing leads employees to view the organization as untrustworthy, opportunistic, and self-serving. The resultant perception of organizational cynicism, as indicated by the research findings, leads to a significant decline in employee morale. This, in turn, impacts efficacy in the workplace, both in terms of task execution and behavior within the organization. Subsequently, the authors do not distinguish between the educational levels of the personnel, as was done in the previously examined study. Li et al. (2022) identify employees' inclination toward a "green" value system as a significant factor that is essential for their response to perceived greenwashing. In a manner akin to the argument presented by Robertson et al. (2023), which is grounded in "social identity theory," Li et al. (2022) employ the "person–organization fit (PO-fit) perspective." This perspective also regards the alignment of values and objectives between the organization and its employees as essential. According to this theory, Li et al. (2022) discovered that a stronger entrenchment of CSR-specific values within the workforce correlates with an increased likelihood of experiencing a person–organization fit conflict in instances of greenwashing. This conflict subsequently fosters a perception of cynicism, which adversely affects employee morale.

Mu and Lee (2023) also investigate the effects of CSR inconsistencies on employees in their study. In contrast to prior research, the authors do not explicitly address the phenomenon of greenwashing; instead, they examine the implications for employees that stem from a preference for external corporate social responsibility (CSR) initiatives over internal ones. Consequently, the emphasis is not explicitly placed on the disparity between rhetoric and action; rather, the study merely addresses the research topic pertinent to this work in a cursory manner. The findings of Mu and Lee (2023) indicate, in alignment with prior research, that inconsistent corporate social responsibility (CSR) strategies are regarded by employees as hypocritical. This perception adversely affects the employer's image and subsequently contributes to an increase in employee turnover. This study elucidates, in alignment with the principles of "social identity theory," that only balanced, authentic, and consistent corporate social responsibility (CSR) strategies enable companies to maintain a favorable reputation over the long term.

References

Aji, H., & Sutikno, B. (2015). The extended consequence of greenwashing: Perceived consumer skepticism. *International Journal of Business and Information., 10*, 433.

Al-Shammari, Z., Faulkner, P. E., & Forlin, C. (2019, June 17). Theories-based inclusive education practices Education Quarterly Reviews, Vol. 2 No. 2, Available at SSRN: https://ssrn.com/abstract=3415247

Amin, A., Chourou, L., Kamal, S., et al. (2020). It's who you know that counts: Board connectedness and CSR performance. *Journal of Corporate Finance, 64*. https://doi.org/10.1016/J.JCORPFIN.2020.101662

Arouri, M., El Ghoul, S., Gomes, M., & Gomes Greenwashing, M. (2021). Greenwashing and product market competition title: Greenwashing and product market competition author names and affiliations: Greenwashing and product market competition. *Finance Research Letters, 42*. https://doi.org/10.1016/j.frl.2021.101927ï

Bais, B., Nassimbeni, G., & Orzes, G. (2024). Global reporting initiative: Literature review and research directions. *Journal of Cleaner Production, 471*. https://doi.org/10.1016/J.JCLEPRO.2024.143428

Behnam, M., & MacLean, T. L. (2011). Where is the accountability in international accountability standards?: A decoupling perspective. *Business Ethics Quarterly, 21*(1), 45–72. Available at SSRN https://ssrn.com/abstract=2019151

Blanco-Davis, E., & Zhou, P. (2016). Life cycle assessment as a complementary utility to regulatory measures of shipping energy efficiency. *Ocean Engineering, 128*, 94–104. https://doi.org/10.1016/j.oceaneng.2016.10.015

Booth, H., Milner-Gulland, E. J., McCormick, N., & Starkey, M. (2024). Operationalizing transformative change for business in the context of nature positive. *One Earth, 7*, 1235–1249. https://doi.org/10.1016/J.ONEEAR.2024.06.003

Bottega, L., Brécard, D., & Delacote, P. (2024). Advertising, ecolabeling and consumers' beliefs: Greenwashing or not? *Economics Letters, 235*. https://doi.org/10.1016/J.ECONLET.2023.111513

Brandão, M., Busch, P., & Kendall, A. (2023). Life cycle assessment, quo vadis? Supporting or deterring greenwashing? A survey of practitioners. *Environmental Science: Advances, 3*, 266–273. https://doi.org/10.1039/D3VA00317E

Brunsson, N. (1993). Ideas and actions: Justification and hypocrisy as alternatives to control. *Accounting, Organizations and Society, 18*, 489–506. https://doi.org/10.1016/0361-3682(93)90001-M

Cahan, S. F., De Villiers, C., Jeter, D. C., et al. (2016). Are CSR disclosures value relevant? Cross-country evidence. *European Accounting Review, 25*, 579–611. https://doi.org/10.1080/09638180.2015.1064009

Cao, J., Faff, R., He, J., & Li, Y. (2022). Who's greenwashing via the media and what are the consequences? Evidence From China. *Abacus, 58*. https://doi.org/10.1111/abac.12273

Chen, P., & Dagestani, A. A. (2023). Greenwashing behavior and firm value – From the perspective of board characteristics. *Corporate Social Responsibility and Environmental Management, 30*, 2330–2343. https://doi.org/10.1002/CSR.2488

Chen, L., Ma, Y., Feng, G. F., & Chang, C. P. (2024). Does environmental governance mitigate the detriment of greenwashing on innovation in China? *Pacific Basin Finance Journal, 86*. https://doi.org/10.1016/J.PACFIN.2024.102450

Chen, Y.-S., & Chang, C.-H. (2013). Greenwash and green trust: The mediation effects of green consumer confusion and green perceived risk. *Journal of Business Ethics, 114*. https://doi.org/10.1007/s10551-012-1360-0

Chin, M. K., Hambrick, D. C., & Treviño, L. K. (2013). Political ideologies of CEOs: The influence of executives' values on corporate social responsibility. *Administrative Science Quarterly, 58*(2), 197–232. https://doi.org/10.1177/0001839213486984. (Original work published 2013).

de Freitas Netto, S. V., Sobral, M. F. F., Ribeiro, A. R. B. et al. (2020). Concepts and forms of greenwashing: a systematic review. *Environmental Sciences Europe 32*,19. https://doi.org/10.1186/s12302-020-0300-3

De Jong, M., Huluba, G., & Beldad, A. (2019). Different shades of greenwashing: Consumers' reactions to environmental lies, half-lies, and organizations taking credit for following legal obligations. *Journal of Business and Technical Communication, 34*, 105065191987410. https://doi.org/10.1177/1050651919874105

Demers, E., Hendrikse, J., Joos, P., & Lev, B. (2021). ESG did not immunize stocks during the COVID-19 crisis, but investments in intangible assets did. *Journal of Business Finance & Accounting, 48*, 433–462. https://doi.org/10.1111/JBFA.12523

Du, X. (2015). How the market values greenwashing? Evidence from China. *Journal of Business Ethics, 128*, 547–574. https://doi.org/10.1007/S10551-014-2122-Y

Ekman, P., & Cordaro, D. (2011). What is meant by calling emotions basic. *Emotion Review, 3*(4), 364–370. https://doi.org/10.1177/1754073911410740

El Koshiry, A., Eliwa, E., Abd El-Hafeez, T., & Shams, M. Y. (2023). Unlocking the power of blockchain in education: An overview of innovations and outcomes. *Blockchain: Research and Applications, 4*. https://doi.org/10.1016/j.bcra.2023.100165

Eliwa, Y., Aboud, A., & Saleh, A. (2021). ESG practices and the cost of debt: Evidence from EU countries. *Critical Perspectives on Accounting, 79*, 102097. https://doi.org/10.1016/J.CPA.2019.102097

Faieq, H. T., & Cek, K. (2024). Enhancing Kurdistan's manufacturing companies' sustainable waste management: A norm activation approach to green accounting, CSR, and environmental auditing oversight. *Heliyon, 10*. https://doi.org/10.1016/j.heliyon.2024.e32725

Fella, S., & Bausa, E. (2024). Green or greenwashed? Examining consumers' ability to identify greenwashing. *Journal of Environmental Psychology, 95*. https://doi.org/10.1016/J.JENVP.2024.102281

García-Lillo, F., Seva-Larrosa, P., & Sánchez-García, E. (2023). What is going on in entrepreneurship research? A bibliometric and SNA analysis. *Journal of Business Research, 158*. https://doi.org/10.1016/j.jbusres.2022.113624

García-Sánchez, I. M., Hussain, N., Aibar-Guzmán, C., & Aibar-Guzmán, B. (2022). Assurance of corporate social responsibility reports: Does it reduce decoupling practices? *Business Ethics, the Environment & Responsibility, 31*, 118–138. https://doi.org/10.1111/BEER.12394

Ghafoor, A., & Gull, A. A. (2024). Do co-opted boards protect CEOs from ESG controversies? *Finance Research Letters, 63*. https://doi.org/10.1016/j.frl.2024.105263

Godos-Diez, J.-L., Cabeza-García, L., & Fernández-González, C. (2018). Relationship between Corporate Social Responsibility (CSR) and internationalisation strategies: A descriptive study in the Spanish Context. *Administrative Sciences, 8*, 57. https://doi.org/10.3390/admsci8040057

Grijalvo, M., & García-Wang, C. (2023). Sustainable business model for climate finance. Key drivers for the commercial banking sector. *Journal of Business Research, 155*. https://doi.org/10.1016/j.jbusres.2022.113446

Gull, A. A., Hussain, N., Khan, S. A., et al. (2023). The power of the CEO and environmental decoupling. *Business Strategy and the Environment, 32*, 3951–3964. https://doi.org/10.1002/BSE.3347

Haque, F., & Ntim, C. G. (2020, January 1). Executive compensation, sustainable compensation policy, carbon performance and market value. *British Journal of Management*, Forthcoming, Available at SSRN: https://ssrn.com/abstract=3512600 or https://doi.org/10.2139/ssrn.3512600

Hawn, O., & Ioannou, I. (2016). Mind the gap: The interplay between external and internal actions in the case of corporate social responsibility. *Strategic Management Journal, 37*, 2569–2588. https://doi.org/10.1002/SMJ.2464

Henao-Rodríguez, C., Lis-Gutiérrez, J. P., & Angulo-Bustinza, H. D. (2024). Unveiling greenwashing in Colombian manufacturing: A machine learning approach. *Research in Globalization, 8*. https://doi.org/10.1016/J.RESGLO.2024.100196

Huang, Y., Xiong, N., & Liu, C. K. (2024). Renewable energy technology innovation and ESG greenwashing: Evidence from supervised machine learning methods using patent text. *Journal of Environmental Management, 370*. https://doi.org/10.1016/j.jenvman.2024.122833

Hussain, T., Wang, D., & Benqian, L. (2023). Examining the role of responsible management, CSR, and TQM in enhancing renewable energy projects: An empirical analysis. *Acta Ecologica Sinica*. https://doi.org/10.1016/j.chnaes.2023.06.010

Huybrechts, B., Pérez-Suárez, M., Cobeña, M., & Sánchez-Torné, I. (2024). Energy co-operatives in Spain: The role of social enterprises in the energy transition. *Futures, 160*. https://doi.org/10.1016/j.futures.2024.103381

Hyatt, D. G., & Berente, N. (2017, December). Substantive or symbolic environmental strategies? Effects of external and internal normative stakeholder pressures. *Business Strategy and the Environment*, Wiley Blackwell, *26*(8), 1212–1234.

Ioannou, I., Kassinis, G., & Papagiannakis, G. (2022). The Impact of Perceived Greenwashing on Customer Satisfaction and the Contingent Role of Capability Reputation. *Journal of Business Ethics, 185*(2), 333–347

Jiang, C., Li, X., Xu, Q., & Liu, J. (2024). Does environmental protection tax impact corporate ESG greenwashing? A quasi-natural experiment in China. *Economic Analysis Policy, 84*, 774–786. https://doi.org/10.1016/J.EAP.2024.09.029

Karagiannis, I., Vouros, P., Sioutas, N., & Evangelinos, K. (2022). Mapping the maritime CSR agenda: A cross-sectoral materiality analysis of sustainability reporting. *Journal of Cleaner Production, 338*. https://doi.org/10.1016/j.jclepro.2021.130139

Kim, E. H., & Lyon, T. P. (2015). Greenwash vs. Brownwash: Exaggeration and undue modesty in corporate sustainability disclosure. *Organization Science, 26*, 705–723. https://doi.org/10.1287/ORSC.2014.0949

Kiran, M., Boubaker, S., Chughtai, S., & Naeem, M. A. (2024). Navigating Greenwashing in the G8: Insights into family-owned firms, technology innovation, and economic policy uncertainty. *Research in International Business and Finance*, 102481. https://doi.org/10.1016/J.RIBAF.2024.102481

Koch, T., & Denner, N. (2025). Different shades of green deception. Greenwashing's adverse effects on corporate image and credibility. *Public Relations Review, 51*. https://doi.org/10.1016/j.pubrev.2024.102521

Koemtzopoulos, D., Zournatzidou, G., & Sariannidis, N. (2025). Can cryptocurrencies be green? The role of stable coins toward a carbon footprint and sustainable ecosystem. *Sustainability, 17*, 483. https://doi.org/10.3390/SU17020483

Li, J. (2024). Controlling shareholders' stock pledges and greenwashing–evidence from China. *Finance Research Letters, 69*. https://doi.org/10.1016/j.frl.2024.106227

Liu, M., Vu, T. V., van Atteveldt, N., & Meeter, M. (2023). Testing the reciprocal effect between value of education, time investment, and academic achievement in a large non-western sample. *Journal of Intelligence, 11*, 1333.

Li, M., & Chen, Q. (2024). Executive pay gap and corporate ESG greenwashing: Evidence from China. *International Review of Financial Analysis, 95.* https://doi.org/10.1016/J.IRFA.2024.103375

Li, W., Li, W., Seppänen V., Koivumäki T. (2023). Effects of greenwashing on financial performance: Moderation through local environmental regulation and media coverage. *Wiley Blackwell, vol. 32*(1), 820–841.

Li, Y., & Xiao, J. (2025). The effect of institutional investors' site visits on corporate greenwashing behavior. *International Review of Economics & Finance, 97*, 103818. https://doi.org/10.1016/J.IREF.2024.103818

Li, D., Zhang, Z., & Gao, X. (2024). Does artificial intelligence deter greenwashing? *Finance Research Letters, 67.* https://doi.org/10.1016/J.FRL.2024.105954

Li, T., Shu, X., & Liao, G. (2024). Does corporate greenwashing affect investors' decisions? *Finance Research Letters, 67.* https://doi.org/10.1016/j.frl.2024.105877

Liu, Y. (2024). Climate risk perception, female directors and greenwashing. *Finance Research Letters, 70*, 106291. https://doi.org/10.1016/J.FRL.2024.106291

Liu, Y., Lei, L., & Buttner, E. H. (2020). Establishing the boundary conditions for female board directors' influence on firm performance through CSR. *Journal of Business Research, 121*, 112–120. https://doi.org/10.1016/J.JBUSRES.2020.08.026

Liu, J.-Y., Lei, Q., Li, R., & Zhang, Y.-J. (2024). Resistance or motivation? Impact of climate risk on corporate greenwashing: An empirical study of Chinese enterprises. *Global Finance Journal*, 101030. https://doi.org/10.1016/J.GFJ.2024.101030

Liu, X., Wang, C., & Wang, X. (2025). Greenwashing and corporate market power. *Journal of Cleaner Production, 486.* https://doi.org/10.1016/j.jclepro.2024.144486

Ma, Y., & Ahmad, M. I. (2024). Do board characteristics impact greenwashing? Moderating role of CSR committee. *Heliyon, 10.* https://doi.org/10.1016/j.heliyon.2024.e38743

Martínez-Ferrero, J., Garcia-Sanchez, I. M., & Cuadrado-Ballesteros, B. (2015). Effect of financial reporting quality on sustainability information disclosure. *Corporate Social Responsibility and Environmental Management, 22*, 45–64. https://doi.org/10.1002/CSR.1330

Miotto, G., Blanco-González, A., & Díez-Martín, F. (2020). Top business schools legitimacy quest through the sustainable development goals. *Heliyon, 6.* https://doi.org/10.1016/j.heliyon.2020.e05395

Montgomery, C., Saini, P., Schoetensack, C., McCarthy, M., Hanlon, C., Owens, L., Kullu, C., van Ginneken, N., Rice, M., & Young, R. (2023a). Improving access to treatment for alcohol dependence in primary care: A qualitative investigation of factors that facilitate and impede treatment access and completion. *PLoS ONE, 18*(10), e0292220. https://doi.org/10.1371/journal.pone.0292220

Mu, H. & Lee, Y. (2023). Greenwashing in Corporate Social Responsibility: A Dual-Faceted Analysis of Its Impact on Employee Trust and Identification. *Sustainability, 15*(22), 15693. https://doi.org/10.3390/su152215693

Nasrullah, N. M. (2011). Business and sustainable development: The role of CSR as a catalyst. *Transnational Corporations Review, 3*, 96–105. https://doi.org/10.1080/19186444.2011.11658300

Nepal, R., Deng, Y., Dong, K., & Goutte, S. (2025). Does executive gender diversity culture inhibit corporate greenwashing behavior? The effect of informal institutions. *International Review of Financial Analysis, 97.* https://doi.org/10.1016/j.irfa.2024.103888

Nerb, J., & Spada, H. (2001). Evaluation of environmental problems: A coherence model of cognition and emotion. *Cognition and Emotion, 15.* https://doi.org/10.1080/02699930143000013

Nyilasy, G., Gangadharbatla, H., & Paladino, A. (2014). Perceived greenwashing: The interactive effects of green advertising and corporate environmental performance on consumer reactions. *Journal of Business Ethics, 125*, 693–707. https://doi.org/10.1007/S10551-013-1944-3

Oguntona, O. A. (2024). Evaluation of global research on greenwashing by scientometric indicators. *Sustainability and Toxicity of Building Materials*, 59–79. https://doi.org/10.1016/B978-0-323-98336-5.00004-2

Paganou, S., Antoniadis, I., Zournatzidou, G., & Sklavos, G. (2024). Investigating the link among corruption, corporate governance and corporate performance in family businesses: A future research agenda. *Administrative Sciences, 14*, 139. https://doi.org/10.3390/ADMSCI14070139

Peng, D., & Kong, Q. (2024). Corporate green innovation under environmental regulation: The role of ESG ratings and greenwashing. *Energy Economics, 140*. https://doi.org/10.1016/j.eneco.2024.107971

Peng, Q., & Xie, Y. (2024). ESG greenwashing and corporate debt financing costs. *Finance Research Letters, 106012*. https://doi.org/10.1016/j.frl.2024.106012

Petrenko, O. V., Aime, F., Ridge, J., & Hill, A. (2016). Corporate social responsibility or CEO narcissism? CSR motivations and organizational performance. *Strategic Management Journal, 37*, 262–279. https://doi.org/10.1002/SMJ.2348

Pizzetti, M., Gatti, L., & Seele, P. (2021). Firms talk, suppliers walk: Analyzing the locus of Greenwashing in the blame game and introducing 'vicarious greenwashing'. *Journal of Business Ethics, 170*, 21–38. https://doi.org/10.1007/S10551-019-04406-2

Ragazou, K., Zournatzidou, G., Sklavos, G., & Sariannidis, N. (2024). Integration of circular economy and urban metabolism for a resilient waste-based sustainable urban environment. *Urban Science, 8*, 175. https://doi.org/10.3390/URBANSCI8040175

Robertson, J., Montgomery, A., & Ozbilir, T. (2023). Employees' response to corporate greenwashing. *Business Strategy and the Environment, 32*, 4015–4027. https://doi.org/10.1002/bse.3351

Ruiz-Blanco, S., Romero, S., & Fernandez-Feijoo, B. (2022). Green, blue or black, but washing–what company characteristics determine greenwashing? *Environment, Development and Sustainability, 24*, 4024–4045. https://doi.org/10.1007/S10668-021-01602-X/TABLES/8

Sauerwald, S., & Su, W. (2019). CEO overconfidence and CSR decoupling. *Corporate Governance: An International Review, 27*. https://doi.org/10.1111/corg.12279

Schons, L.,& Steinmeier, M., (2016). Walk the Talk? How Symbolic and Substantive CSR Actions Affect Firm Performance Depending on Stakeholder Proximit. *Corporate Social Responsibility and Environmental Management, 23*(6), 358–372

Seele, P., & Gatti, L. (2017, February). Greenwashing revisited: In search of a typology and accusation-based definition incorporating legitimacy strategies. *Business Strategy and the Environment*, Wiley Blackwell, *26*(2), 239–252.

Sun, H., Bai, T., & Liu, Z. (2024). Perceived policy uncertainty and enterprise greenwashing risk. *Finance Research Letters, 69*. https://doi.org/10.1016/j.frl.2024.106148

Szabo, S., & Webster, J. (2021, July). Perceived greenwashing: The effects of green marketing on environmental and product perceptions. *Journal of Business Ethics*, Springer, *171*(4), 719–739.

Tao, Z. (2025). From green demand to deceptive practices: A cross-cultural study the impact of consumer green preferences on corporate greenwashing behavior in China, Republic of Korea, and Malaysia. *Sustainable Futures, 9*. https://doi.org/10.1016/j.sftr.2025.100440

Testa, F., Miroshnychenko, I., Barontini, R., & Frey, M. (2018). Does it pay to be a greenwasher or a brownwasher? *Business Strategy and the Environment, 27*, 1104–1116. https://doi.org/10.1002/BSE.2058

Testa, F., Boiral, O., & Iraldo, F. (2018). Internalization of environmental practices and institutional complexity: Can stakeholders pressures encourage greenwashing? *Journal of Business Ethics, 147*, 287–307. https://doi.org/10.1007/S10551-015-2960-2

Teti, E., Etro, L. L., & Pausini, L. (2024). Does greenwashing affect company's stock Price? Evidence from Europe. *International Review of Financial Analysis, 93*. https://doi.org/10.1016/J.IRFA.2024.103195

Tong, Y., Lau, Y. W., & Binti Ngalim, S. M. (2024). Do pilot zones for green finance reform and innovation avoid ESG greenwashing? Evidence from China. *Heliyon, 10*. https://doi.org/10.1016/j.heliyon.2024.e33710

Torelli, R., Balluchi, F., & Lazzini, A. (2020). Greenwashing and environmental communication: Effects on stakeholders' perceptions. *Business Strategy and the Environment, 29*, 407–421. https://doi.org/10.1002/BSE.2373

Tsukada, A., Okamatsu-Ogura, Y., Futagawa, E., et al. (2023). White adipose tissue undergoes browning during preweaning period in association with microbiota formation in mice. *iScience, 26*(7), 107239. https://doi.org/10.1016/J.ISCI.2023.107239

Tzitiridou-Chatzopoulou, M., & Zournatzidou, G. (2024). Bibliometric analysis on of the impact of screening to minimize maternal mental health on neonatal outcomes: A systematic review. *Journal of Clinical Medicine, 13*, 6013. https://doi.org/10.3390/JCM13196013

Tzitiridou-Chatzopoulou, M., Orovou, E., & Zournatzidou, G. (2024). Digital training for nurses and midwives to improve treatment for women with postpartum depression and protect neonates: A dynamic bibliometric review analysis. *Healthcare, 12*, 1015. https://doi.org/10.3390/HEALTHCARE12101015

Tzitiridou-Chatzopoulou, M., Zournatzidou, G., Tsakiridis, I., & Tsakalidis, C. (2024). Exploring the nexus between fertility rates and geopolitical risk with intelligence methods: A multifaceted analysis. *Healthcare, 12*, 2205. https://doi.org/10.3390/HEALTHCARE12222205

van Oorschot, K. E., Aas Johansen, V., Lynes Thorup, N., & Aspen, D. M. (2024). Standardization cycles in sustainability reporting within the global reporting initiative. *European Management Journal*. https://doi.org/10.1016/J.EMJ.2024.04.001

Vázquez-Ingelmo, A., García-Holgado, A., Verdugo-Castro, S., et al. (2024). Data visualization and domain knowledge: Insights through focus groups of researchers in Spain. *Computers in Human Behavior, 155*. https://doi.org/10.1016/j.chb.2024.108162

Walker, K., & Wan, F. (2012). The harm of symbolic actions and green-washing: Corporate actions and communications on environmental performance and their financial implications. *Journal of Business Ethics, 109*. https://doi.org/10.1007/s10551-011-1122-4

Wang, Z., & Sarkis, J. (2017). Corporate social responsibility governance, outcomes, and financial performance. *Journal of Cleaner Production, 162*, 1607–1616. https://doi.org/10.1016/J.JCLEPRO.2017.06.142

Wang, H., & Shen, H. (2024). Self-restraint or external supervision: Green bond issuance and greenwashing. *Research in International Business and Finance, 70*. https://doi.org/10.1016/J.RIBAF.2024.102402

Wang, Z., Hsieh, T. S., & Sarkis, J. (2018). CSR performance and the readability of CSR reports: Too good to be true? *Corporate Social Responsibility and Environmental Management, 25*, 66–79. https://doi.org/10.1002/CSR.1440

Wang, Q., Ma, Z., Zhao, J., & Shu, G. (2024). State shareholding in privately-owned firms and greenwashing: Evidence from China. *Finance Research Letters, 62*. https://doi.org/10.1016/J.FRL.2024.105176

Wang, Y., Xing, C., & Zhang, L. (2024). Is greenwashing beneficial for corporate access to financing? Evidence from China. *Finance Research Letters, 65*. https://doi.org/10.1016/J.FRL.2024.105538

Xia, F., Chen, J., Yang, X., et al. (2023). Financial constraints and corporate greenwashing strategies in China. *Corporate Social Responsibility and Environmental Management, 30*, 1770–1781. https://doi.org/10.1002/CSR.2453

Xu, T., Sun, Y., & He, W. (2024). Government digitalization and corporate greenwashing. *Journal of Cleaner Production, 452*. https://doi.org/10.1016/J.JCLEPRO.2024.142015

Yang, Z., Nguyen, T. T. H., Nguyen, H. N., Nguyen, T. T. N., & Cao, T. T. (2020). Greenwashing behaviours: Causes, taxonomy and consequences based on a systematic literature review. *Journal of Business Economics and Management, 21*(5), 1486–1507. https://doi.org/10.3846/jbem.2020.13225

Yao, K., Ma, X., & Zhang, J. (2024). Green disguise or real action? The truth behind corporate greenwashing under stringent climate policy. *Energy Economics*, 108059. https://doi.org/10.1016/J.ENECO.2024.108059

Yu, E. P., Van Luu, B., & Chen, C. H. (2020). Greenwashing in environmental, social and governance disclosures. *Research in International Business and Finance, 52*, 101192. https://doi.org/10.1016/J.RIBAF.2020.101192

Yu, X., Xiao, K., & Xu, T. (2023). Does ESG profile depicted in CSR reports affect stock returns? Evidence from China. *Physica A: Statistical Mechanics and its Applications, 627*. https://doi.org/10.1016/j.physa.2023.129118

Zhang, D. (2022). Are firms motivated to greenwash by financial constraints? Evidence from global firms' data. *Journal of International Financial Management & Accounting, 33*, 459–479. https://doi.org/10.1111/JIFM.12153

Zhao, X., Huang, X., Liu, F., & Pan, L. (2024). Executive power discrepancy and corporate ESG greenwashing. *International Review of Financial Analysis, 96*. https://doi.org/10.1016/j.irfa.2024.103533

Zhou, K., Qu, Z., Liang, J., et al. (2024). Threat or shield: Environmental administrative penalties and corporate greenwashing. *Finance Research Letters, 61*. https://doi.org/10.1016/J.FRL.2024.105031

Zournatzidou, G. (2024). Evaluating executives and non-executives' impact toward ESG performance in banking sector: An entropy weight and TOPSIS method. *Administrative Sciences, 14*, 255. https://doi.org/10.3390/ADMSCI14100255

Zournatzidou, G., & Floros, C. (2023). Hurst exponent analysis: Evidence from volatility indices and the volatility of volatility indices. *Journal of Risk and Financial Management, 16*, 272. https://doi.org/10.3390/JRFM16050272

Zournatzidou, G., Sklavos, G., Ragazou, K., & Sariannidis, N. (2024a). Anti-competition and anti-corruption controversies in the European financial sector: Examining the anti-ESG factors with entropy weight and TOPSIS methods. *Journal of Risk and Financial Management, 17*, 492. https://doi.org/10.3390/JRFM17110492

Zournatzidou, G., Mallidis, I., Farazakis, D., & Floros, C. (2024). Enhancing bitcoin price volatility estimator predictions: A four-step methodological approach utilizing elastic net regression. *Mathematics, 12*, 1392. https://doi.org/10.3390/MATH12091392

Zournatzidou, G., Ragazou, K., Sklavos, G., et al. (2024a). (1AD) The influence of greenwashing methods on the behavior of environmentally conscious employees. In R. Konstantina, G. Alexandros, P. Christos, & S. Angeliki (Eds.), *Sustainability through green HRM and performance integration* (pp. 39–58). IGI Global.

Zournatzidou, G., Ragazou, K., Samantzis, C., et al. (2024). (1AD) Looking back and looking forward: Current and future practices in green human resources management (RM) and green innovation. In R. Konstantina, G. Alexandros, P. Christos, & S. Angeliki (Eds.), *Sustainability through green HRM and performance integration* (pp. 441–458). IGI Global.

Zournatzidou, G., Ragazou, K., Sklavos, G., & Sariannidis, N. (2024). (1AD) Linking HRM with sustainability: Achieving the goals of the 2030 agenda. In R. Konstantina, G. Alexandros, P. Christos, & S. Angeliki (Eds.), *Sustainability through green HRM and performance integration* (pp. 281–304). IGI Global.

Zournatzidou, G., Sklavos, G., Ragazou, K., & Sariannidis, N. (2024b). (1AD) Building carbon-neutral green HRM resources through the use of green financing. In R. Konstantina, G. Alexandros, P. Christos, & S. Angeliki (Eds.), *Sustainability through green HRM and performance integration* (pp. 219–242). IGI Global.

Zournatzidou, G., Farazakis, D., Sklavos, G., et al. (2024). *(1AD) Sustainability, green financing, and corporate management*. https://services.igi-global.com/resolvedoi/resolve.aspx?doi=104018/979-8-3693-5981-5.ch009191–218. IGI Global. https://doi.org/10.4018/979-8-3693-5981-5.CH009

Zournatzidou, G., Ragazou, K., Sklavos, G., et al. (2024b). (1AD) Digital transformation and sustainable HRM: The challenges for an eco-friendly business continuity. In R. Konstantina, G. Alexandros, P. Christos, & S. Angeliki (Eds.), *Sustainability through green HRM and performance integration* (pp. 421–440). IGI Global.

Greenwashing Versus Brownwashing: The Dichotomy of Exaggeration and Undue Modesty in Corporate Sustainability Practices: Disclosure

2

2.1 Definition of Brownwashing

Brownwashing can be characterized as a form of "shadow strategy" in which corporations portray their actual practices concerning social and environmental issues as being less significant than they truly are. In the subsequent section of the master's thesis, an analysis is conducted regarding the motivations that may lead companies to downplay their image of social responsibility or conceal their commitments entirely, rather than transparently communicating their endeavors in these areas. Furthermore, an examination was conducted on the factors that may mitigate or inhibit the occurrence of brownwashing.

2.1.1 Extent of Deregulation Within the Business Environment

In a business environment characterized by stringent regulatory oversight, there exist explicit regulations and statutes that dictate the conduct of corporations and their executives. Through regulatory measures, the state plays an active role in ensuring that companies operate in alignment with the public interest, particularly with regard to the stakeholders involved. An illustrative instance of this phenomenon is the market for electric mobility within the United States of America. This is subject to rigorous oversight by the Federal Energy Regulatory Commission, the State Public Utility Commission, and the Securities and Exchange Commission. These authorities have a strong influence on the business activities of companies active in the electric vehicle construction sector, for example, by setting quotas for the return on capital investment but also by requiring regulations for the disclosure of certain CSR standards. Consequently, companies are compelled by governmental regulations to adhere to stringent financial criteria and environmental protection measures. This, in theory, implies that they possess no latitude to engage in

K. Ragazou et al., *Transparency and Corporate Washing in the Age of Sustainability*, Management for Professionals,
https://doi.org/10.1007/978-3-031-96821-1_2

opportunistic behavior that would serve their own interests or those of their shareholders. According to the principal–agent theory, the state minimizes the conflict between shareholders and companies through extant regulations, which means that companies have little capacity for opportunistic behavior, such as brownwashing. In instances where an organization functions within an environment characterized by minimal or nonexistent governmental regulation, shareholders exhibit heightened scrutiny toward the company. Corporate Social Responsibility (CSR) investments are frequently perceived as cost liabilities that do not generate immediate financial returns. To achieve the profitability objectives set forth by shareholders, companies may consider diminishing their expenditures on environmental protection and social benefits, as investors often perceive these initiatives as a financial burden. This indicates that, in the absence of governmental regulation, there exists a risk that companies may underreport their disclosed expenditures on Corporate Social Responsibility (CSR) in relation to their actual investments. Engaging in brownwashing behavior, however, engenders a principal–agent conflict between corporations and their shareholders.

Kim and Lyon (2015) investigate this phenomenon and conclude that, within a deregulated environment, shareholders apply pressure on the corporation. In order to maintain favorable relations with their investors, companies feel obligated to diminish their corporate social responsibility (CSR) performance, at least in documentation. Consequently, a deregulated corporate environment results in heightened scrutiny and pressure from shareholders, which explains why organizations often downplay their substantial investments in corporate social responsibility (CSR). Moreover, Kim and Lyon (2015) discovered evidence indicating that, in accordance with the principal–agent theory, in a deregulated corporate milieu characterized by low and diminishing profitability, shareholders assume the predominant role of stakeholders. This is evidenced by the observation that investors, perceiving inadequate profitability, harbor suspicions of opportunistic behavior, consequently exerting pressure on the company. The research indicates that a net revenue decrease of one trillion US dollars correlates with a 15.7% increase in the probability of brownwashing. It can be inferred that when a company's profitability fails to align with investors' expectations, the organization is necessitated to prioritize short-term gains in order to satisfy shareholder demands. Consequently, investments in Corporate Social Responsibility (CSR), which yield long-term profitability, are initially obscured from investors.

2.1.2 Legitimacy of the Company

The likelihood of a company engaging in brownwashing may be significantly influenced by the legitimacy status attributed to that company. According to legitimacy theory, an organization is deemed legitimate when the pertinent constituent groups regard its actions as desirable or appropriate. The literature delineates three distinct types. Pragmatic legitimacy addresses the economic interests of stakeholders, whereas moral legitimacy emphasizes actions that promote the collective welfare.

Ultimately, cognitive legitimacy is established when the existence and objectives of the company are deemed indispensable. In this context, there is a suspicion that notable companies with high legitimacy conceal their CSR activities from their stakeholders. Huang et al. (2024) examine this phenomenon and, based on their analysis, conclude that firms recognized for their exceptional environmental performance may be more inclined to engage in brownwashing practices. The authors contend that even when a company significantly excels in its contributions to environmental protection relative to its peers, the organization modifies its public image to conform with industry standards and conceals any supplementary initiatives. The fundamental rationale is that by obscuring corporate social responsibility (CSR) accomplishments, the overall standard within the peer group remains unenhanced. The company anticipates that stakeholders will refrain from imposing any new demands, thereby preserving its status of legitimacy. In this context, the elevated degree of legitimacy posited by legitimacy theory plays a substantial role in the phenomenon of brownwashing. Conversely, organizations that modify their publication practices to align with those of their peer group may be culpable of the antithesis of brownwashing, specifically, greenwashing.

2.1.3 Compensation of Stakeholders

The chapter posits that when a corporation adequately attends to a stakeholder group, it engenders favorable responses and perceptions from all parties involved. An illustrative example of this would be a firm choosing to participate in philanthropy by offering financial assistance to a disadvantaged group. The contributing corporation anticipates acquiring a favorable reputation from the recipient organization and its stakeholder groups, although it overlooks the possible risk of dispute that may ensue. This risk delineates the possible conflict between the company's social responsibility and its pragmatic credibility among stakeholders. For a more detailed explanation, it is prudent to examine Carroll's CSR pyramid. Economic prosperity underpins the main stakeholders, including workers and investors. This stakeholder group deems the company's social and/or sustainable engagement acceptable only when the economic component is seen as sufficient.

2.1.3.1 Stakeholders with a Short-Term Horizon

When considering employees as a distinct interest group, remuneration emerges as the pivotal criterion for the pragmatic legitimacy of the employer. Compensation is a critical determinant of employees' perceptions regarding their value and equitable treatment within the organization. The appropriateness of remuneration is assessed internally through informal discussions among individual employees or externally by comparing salary disparities among companies operating within the same industry. When employees perceive their remuneration as inequitable, it results, among other consequences, in diminished motivation and productivity within the workplace, as well as an increase in employee disloyalty, leading to a greater propensity to seek alternative employment opportunities. When evaluating pragmatic legitimacy from the standpoint of investors, it is imperative that the investment yield a return that meets, or preferably surpasses, the investors' expectations. Should this criterion be satisfied, investors are inclined to diverge from the neoclassical finance theory and endorse social and sustainable initiatives undertaken by companies.

Should investors' expectations remain unfulfilled, they typically respond by divesting their investments or exerting pressure on the management. The employer may attempt to articulate justifications such as resource scarcity, investments aimed at future growth, or the potentially suboptimal current performance of the business model to foster comprehension regarding the inadequacy of the remuneration in relation to market standards. Should these justifications appear credible to the employee or investor, the company's legitimacy may be upheld notwithstanding diminished earnings. The employee's reasoned inference, however, indicates that this also implies a reduction in the company's resources allocated for corporate social responsibility or philanthropic endeavors. When an organization allocates financial resources to these causes, despite having undercompensated employees or investors, its legitimacy diminishes. In response to this situation, the company may opt to conceal its philanthropic and corporate social responsibility (CSR) initiatives from the public, particularly its employees, while potentially engaging in practices that could be characterized as brownwashing. Miao et al. (2023) present evidence of brownwashing in their examination of publicly traded Chinese corporations. Organizations that compensate their employees or investors at a rate below the average fail to achieve even the most fundamental tier of economic success, as delineated in Carroll's Corporate Social Responsibility (CSR) pyramid. In this regard, it is unjust for employees and investors if the company engages in social and sustainability matters, as such involvement is viewed as inequitable. To mitigate potential conflicts among corporations, employees, and investors, Miao et al. (2023) observe that Chinese companies tend to underreport their philanthropic contributions in their published materials. This suggests that the compensation of employees and investors, as well as the overall legitimacy of the company, serves as a critical factor in the phenomenon of brownwashing.

2.1.3.2 Stakeholders with a Long-Term Horizon

The prior perspective on pragmatic legitimacy suggests that all stakeholders engage solely in short-term actions driven by profit motives. When examining social and environmental expenditures within a short-term time frame, it becomes evident that these represent a considerable cost factor. An examination of Corporate Social Responsibility (CSR) expenditures over an extended timeframe reveals a positive correlation with a company's financial performance. In light of this, it is conceivable that employees or investors may acquiesce to a comparatively modest salary or return in the interim while simultaneously endorsing social and environmental initiatives, which stands in contrast to the assertions made in the preceding sections. This suggests that an approach that considers solely short-term-oriented stakeholders is excessively narrow in scope. A pertinent indicator of a company's long-term orientation is the implementation of "employee incentive programs." Organizations that implement employee retention programs seek to cultivate a robust and enduring relationship with their workforce, thereby fostering loyalty and establishing legitimacy. Examples of this practice encompass the issuance of employee shares, the provision of financial incentives for prolonged employment, or the distribution of compensation over an extended period. This indicates that employees are invested in the long-term viability and prosperity of the company and, conversely, perceive expenditures on Corporate Social Responsibility (CSR) as justifiable. Consequently, corporations possess no motivation to obscure their social and environmental advantages, thereby mitigating the occurrence of brownwashing.

Frequently, a company need not enhance its appeal through financial initiatives. The literature provides evidence indicating that employees and investors exhibit a pronounced sense of affiliation with companies that share their values. Consequently, employees derive satisfaction from their tasks, leading to elevated workplace morale and rendering resignation a relatively improbable occurrence. Consequently, organizations consistently prioritize the recruitment of employees whose values and dispositions align with those of the organization. In relation to a company's long-term orientation, this indicates that the organization specifically recruits employees who align with this strategic approach and are therefore motivated to contribute to the company's enduring success. This indicates that employees comprehend the expenses related to Corporate Social Responsibility (CSR) and are willing to tolerate reduced compensation in the short term to facilitate long-term sustainability. By comprehending the interests of its stakeholders, the company is not inclined to conceal information; rather, it is able to engage in transparent communication. Consequently, there exists no justification for the practice of brownwashing.

On the part of investors, it is important to note that not all individuals are solely focused on the company's short-term achievements. A crucial differentiation is established between institutional investors and private investors. Institutional investors typically possess substantial holdings of shares, thereby exerting significant influence over corporate governance. In contrast to private investors, their investment strategy is oriented toward the long term, indicating that short-term losses and insufficient dividends do not constitute determining factors for the prompt divestiture of shares. Conversely, there are assertions that investments in Corporate Social Responsibility (CSR) are perceived favorably, as they are regarded as long-term investments in the organization, thereby enhancing the company's value in the future. Miao et al. (2023) provide evidence indicating that stakeholders with a long-term orientation are more likely to possess a comprehensive understanding of a company's social initiatives compared to their short-term-oriented counterparts, even in instances where economic requirements remain unfulfilled. Companies that receive investments from institutional investors are unlikely to obscure their corporate social responsibility (CSR) initiatives, even in instances where the returns for investors may be deemed inadequate. Taking into account the specified time horizon, the compensation of stakeholders can consequently be regarded as a pivotal factor influencing the phenomenon of brownwashing.

2.1.4 Debt of a Company

According to legitimacy theory, actions that contravene societal values and norms may undermine investor trust. This erosion of trust can consequently restrict access to new capital and lead to adverse effects on a company's capital acquisition costs. Organizations endeavor to cultivate an image that is both socially and environmentally responsible. In alignment with this, Zhang (2022) and Xia et al. (2023) discovered in their respective studies that corporations participate in greenwashing as a means to enhance their image as environmentally responsible and socially committed entities. Considering that a socially and environmentally responsible image positively influences capital acquisition, it can be posited that companies experiencing financial difficulties are particularly unlikely to downplay their corporate social responsibility (CSR) initiatives. This argument posits that brownwashing would result in the company intentionally subjecting itself to elevated financing costs, thereby deteriorating its competitive standing. Huang and Kung (2010) conducted an investigation utilizing data from companies listed on the Taiwan Stock Exchange and identified a substantially positive correlation between a company's debt levels and its sustainability disclosures. The authors substantiate this assertion by positing that elevated levels of debt correspond to an increased degree of risk. Should a company fail to conduct itself appropriately in relation to corporate social responsibility (CSR), it may face financial repercussions, potentially hindering its ability to service its loan obligations. Debtors are required to transparently disclose their Corporate Social Responsibility (CSR) activities, enabling creditors to implement appropriate countermeasures in the event of any uncertainties. Research conducted by Huang et al. (2024), which was undertaken a few years later, indicates that elevated corporate debt constrains brownwashing practices, as financiers within this context require a greater degree of information density.

According to legitimacy theory, a company's image and reputation may be adversely affected by the cautious and less transparent reporting of corporate social responsibility (CSR) initiatives. Consequently, this would result in the company encountering challenges in securing capital. Specifically, organizations that are already experiencing financial distress are unlikely to engage in brownwashing, as indicated by the research conducted by Huang et al. (2025). A significant level of indebtedness, coupled with the necessity for investors to receive accurate and efficient reporting, along with the quest for legitimacy, compels companies to diminish their engagement in brownwashing practices.

2.1.5 Supply Chain Management

Within the context of a company's supply chain, both upstream and downstream processes constitute a fundamental component. In addition to economic considerations, the image and reputation of a corporation play a significant role in determining whether an organization will be integrated into this supply chain. In this context, it is imperative to implement measures that enhance the public perception of the company, thereby strengthening its competitive standing. One effective method to enhance the company's reputation is through the implementation of corporate social responsibility (CSR) initiatives, including but not limited to "corporate volunteering" activities. Huang and Kung (2010) identified a positive correlation between corporate publications that highlight a company's corporate social responsibility (CSR) performance and the establishment of a favorable corporate image. In light of the findings presented in the previous subsection, it can be inferred that competitive pressure within the supply chain may inhibit companies from obscuring their social and environmental performance from public scrutiny. Conversely, the benefits of transparent reporting would likely enhance the company's reputation, thereby fostering potential collaboration as a supplier. Negri et al. (2021) assert in their literature review that "sustainable supply chain management," defined as the incorporation of economic, social, and environmental considerations into the supply chain decision-making process, is fundamental to securing enduring corporate performance and mitigating risks. In this context, competitive pressures and the concomitant necessity for legitimacy within the supply chain may serve to mitigate the occurrence of brownwashing.

However, competitive pressure within the supply chain is not the sole factor that can be recognized as a constraint on brownwashing. At present, the introduction of the EU Supply Chain Act is under deliberation within the European Union (EU) in relation to the current status of this master's thesis. In the future, this directive aims to establish regulations governing the due diligence responsibilities of corporations in relation to adherence to human rights and the safeguarding of environmental protection. A supply chain, often referred to as a supply chain, encompasses all activities, resources, and organizations engaged in the production and delivery of a product or service. A differentiation is established between the upstream and downstream process stages and organizations. The former pertains, for instance, to the acquisition of basic materials and the ensuing production processes. Downstream entities or organizations pertain, for instance, to subsequent refining or distribution activities. The supply chain legislation mandated by the European Union now obligates companies to systematically monitor and report all effects pertaining to human rights and environmental considerations throughout the entirety of the value chain. Ultimately, organizations ought to compile a risk management report that delineates the adverse effects on human rights and the environment while also articulating the measures undertaken by the company to mitigate or avert such impacts. An EU directive that delineates explicit requirements and stipulates penalties for non-compliance could impose stringent implications for corporate transparency obligations and the reporting of corporate social responsibility (CSR). An obligation on the part of the legislator, such as the EU supply chain law, would force companies to disclose their social and environmental impacts and thus represent a far-reaching limitation on brownwashing.

2.1.6 Size of the Company

In comparison to smaller enterprises, larger corporations typically possess a more prominent public profile, often attributable to their marketing initiatives as well as the nature of their products or services. This indicates that large corporations are subject to heightened scrutiny by the media and bear a more significant social responsibility. This is closely aligned with the "Political Cost Hypothesis." This indicates that large organizations, in particular, are subject to heightened scrutiny from the public owing to their significant influence on the economy and society, which may result in potential political challenges for these companies. The obstacles are manifested through governmental interventions that, for instance, compel large corporations to adhere to heightened transparency requirements. Nevertheless, global entities are not solely accountable for oversight by the public or governmental authorities. Internal constituents, particularly employees, exhibit a vested interest in adherence to social and environmental considerations. Employees, along with works councils and unions, exert pressure on companies regarding these issues.

In relation to company size, Brammer and Pavelin (2006), along with Li Huang and Kung (2010), observe that larger enterprises are subjected to heightened scrutiny from both internal and external stakeholders. This increased oversight suggests that such companies are likely to engage in more extensive corporate social responsibility (CSR) reporting, even on a voluntary basis. Global entities are not perceived to be engaging in brownwashing in this context. Prior to directing attention to small and medium enterprises (SMEs) in the subsequent section, it is essential to elucidate the definition of SMEs in this context. When taking into account the recommendations of the Austrian Chamber of Commerce, the criterion of having fewer than 250 employees remains a significant factor. Furthermore, it is stipulated that a turnover threshold of EUR 50 million, a total balance sheet amount of EUR 43 million, and a capital share or voting rights of no less than 75% should not be publicly traded. Nonetheless, this solely constitutes a recommendation from the Chamber of Commerce regarding the classification of companies based on their scale. There exists no definitive characterization for small and medium enterprises. Upon examining these figures, it is evident that small and medium enterprises (SMEs) are significantly disadvantaged compared to global corporations in terms of both human resources and financial capital. Conversely, this does not necessarily imply that small and medium enterprises (SMEs) are uninvolved in social and environmental considerations. Conversely, the literature frequently indicates that smaller enterprises demonstrate a commitment to corporate social responsibility (CSR); however, they typically do not publicize this commitment through reports or on their websites. This hesitance to disseminate information may stem from a variety of factors.

A potential initial framework for this can be found in the "Resource-Based View." This asserts that the organization's resources and capabilities are finite and must be utilized with optimal efficiency. An illustration of constrained resources is the establishment of a distinct Environmental, Social, and Governance (ESG) entity, tasked with the preparation and promotion of the company's Corporate Social

Responsibility (CSR) reports and initiatives. In the context of large corporations, this department merely constitutes a small portion of the overall expenses. However, if one looks at SMEs, the cost burden induced by an ESG entity represents a far more significant component compared to the expenditure on the primary business. In the context of internal communication pertaining to a Corporate Social Responsibility (CSR) strategy, smaller enterprises may not require a dedicated Environmental, Social, and Governance (ESG) unit to effectively convey the project to their employees in a transparent manner. The implementation of social and ecological considerations within organizational processes is facilitated by reduced hierarchical levels and a generally diminished degree of bureaucracy. The initiatives may be disseminated through informal communication within the organization. The external reporting of activities, whether in accordance with the Global Reporting Initiative (GRI) or the Non-Financial Reporting Initiative, is typically mandated only for organizations with a specified number of employees and is accompanied by a financial burden. Owner-managed companies are also only minimally dependent on any shareholders, which means that there is no pressure from these interest groups to publish. Instead, informal face-to-face interactions are employed with a select group of pertinent shareholders, rather than relying on a formal Corporate Social Responsibility (CSR) report for disseminating information. The "resource-based view" suggests that small and medium enterprises (SMEs) have minimal necessity to disseminate their social and environmental commitments through media channels. Conversely, it appears that relatively fewer enterprises are more inclined to engage in brownwashing as a result of this phenomenon (Ruiz-Blanco et al., 2022; Trim & Jones, 2025).

2.1.7 Non-Governmental Organizations

Non-governmental organizations (NGOs) are entities that operate independently of governmental influence and are dedicated to promoting the common welfare without being swayed by political or economic interests. In contrast to oversight by shareholders, who prioritize profitability among various factors, companies are purportedly less predisposed to engage in brownwashing when subjected to scrutiny by non-governmental organizations (NGOs). This is attributable to the perception among shareholders that investments in Corporate Social Responsibility (CSR) represent an expense that does not yield an immediate return. Consequently, investors may regard Corporate Social Responsibility (CSR) initiatives with skepticism, potentially prompting companies to obscure such efforts. Conversely, governmental institutions or non-governmental organizations may perceive declining profits as less significant. The rationale underlying this assertion is that a decline in a company's profitability corresponds with a reduction in its ecological legacy. Consequently, this implies that within a regulated business environment, a decline in profitability would eliminate any incentive for companies to partake in brownwashing practices. Brownwashing refers to the practice whereby a corporation portrays itself as having inferior environmental performance compared to its actual standing (Bernard & Nicolau, 2022; Chen & Dagestani, 2023; Qiao et al., 2025; Torelli et al., 2020; Ye et al., 2024).

This situation would render the organization vulnerable to heightened hostility from non-governmental organizations (NGOs) or the media due to insufficient social and environmental performance, compounding an already precarious economic position. Kim and Lyon (2015) demonstrate that in a corporate setting devoid of the involvement of a non-governmental organization (NGO), a decline in net profits by one trillion US dollars correlates with an increase in the likelihood of brownwashing by 188.1%. Should net profits decline by one trillion US dollars in the presence of a non-governmental organization (NGO), the likelihood of underreported social and environmental performance would rise by 133.4%. Specifically, Kim and Lyon (2015) use the Sierra Club as an NGO. The results presented in Sect. 3.1.1, along with the study conducted by Kim and Lyon (2015) and the research undertaken by Montgomery et al. (2023b), elucidate the impact of monitoring on corporate policy and, subsequently, on the phenomenon of brownwashing (Strand & Freeman, 2015; Testa et al., 2018).

2.1.8 Environmentally Sensitive Industries

In the section addressing greenwashing within this volume (Chap. 1), it has been previously articulated, drawing upon the research conducted by Marquis et al. (2016) and Ruiz-Blanco et al. (2022), that corporations engaged in environmentally sensitive sectors may encounter heightened scrutiny, oversight, and consequently, pressure from stakeholders. In reference to the non-governmental organizations previously mentioned, Halme and Hause identified as early as 1977 that these organizations focus their efforts on industries that exert a significant environmental impact. Huang and Kung (2010) assert that stakeholder monitoring, along with the influence exerted by non-governmental organizations (NGOs), can significantly impact corporate management practices. This influence is manifested in the heightened publication output of those companies that are prone to environmental pollution. The study indicates that particularly lucrative enterprises within sectors such as petroleum, chemicals, metals, and paper are increasingly disseminating Corporate Social Responsibility (CSR) reports. Concerning both greenwashing and brownwashing, the literature reveals consistent findings indicating that heightened pressure and the pursuit of legitimacy result in more accurate reporting, as posited by the legitimacy theory. Consequently, this leads to a purportedly reduced discrepancy between rhetoric and action (Hussain et al., 2023; Petrenko et al., 2016; Wang et al., 2018; Yu et al., 2023).

2.1.9 Corporate Governance Mechanism

The phrase "corporate governance management" pertains to the economic, social, and sustainable practices implemented by the organization. The objective of effectively managing these practices is to diminish the information asymmetry that exists between the organization and its stakeholders. In the literature, corporate governance mechanisms that achieve these objectives while simultaneously safeguarding the environment are designated as "sound corporate governance practices." The existing literature indicates that the composition of the board plays a significant role in the execution of social and sustainable practices. It is evident that organizations characterized by a diverse structural composition, a substantial number of board members, boards that convene across multiple enterprises, and those with partial state ownership are more inclined to incorporate Corporate Social Responsibility (CSR) strategies into their operational practices (Mandal & Murthy, 2021; Zhang et al. 2024).

Nevertheless, corporate governance mechanisms assume a critical role in organizations that lack these specific board attributes. These facilitate the development of a comprehension among employees and administrators regarding environmentally sustainable management decisions and their execution within the organization. Ultimately, it is the management that determines the corporate social responsibility (CSR) practices to be implemented by the company and subsequently communicates these initiatives to its stakeholders. Nevertheless, research indicates that enhancing awareness of this issue both internally and externally can lead to a reduction in the economic footprint of certain companies while simultaneously increasing the quality and quantity of publications pertaining to the subject of Corporate Social Responsibility (CSR). In this context, the prevailing theory posits that the effective implementation of corporate governance mechanisms—such as a corporate strategy that incorporates corporate social responsibility (CSR) initiatives, an environmental management system (EMS), or a dedicated CSR committee—has the potential to mitigate instances of brownwashing (Gacek, 2020; Lagoarde-Ségot, 2024; Li et al. 2024; Wang et al. 2024; Wang & Shen, 2024, Zhang & He, 2025).

2.1.9.1 Integrated CSR Strategy

To foster awareness of socially responsible and sustainable corporate governance, it is imperative to incorporate Corporate Social Responsibility (CSR) elements into the existing business strategy. The objective is to incorporate the environmental considerations of stakeholders, which have historically been somewhat variable, into the fundamental operations of the organization. It is essential to achieve a harmonious equilibrium among the economic, social, and environmental dimensions. It is important to acknowledge that the incorporation of a Corporate Social Responsibility (CSR) strategy into business operations is not a singular initiative. Instead, it constitutes a continuous and dynamic process in which inefficiencies on both economic and ecological fronts must consistently be identified and, when required, realigned with the overarching objectives. In their research, Garcia-Sanchez et al. (2021) underscore the beneficial elements associated with the incorporation of Corporate Social Responsibility (CSR) into corporate governance, notably the enhancement of internal communication among managers. Huang et al. (2024) also state that the integration of CSR into the corporate strategy is an essential approach to mitigating brownwashing (Katsaros et al., 2024; Zhang et al. 2024).

2.1.9.2 Environmental Management System

The objective of an Environmental Management System (EMS) is to assist organizations in the implementation, evaluation, and reporting of sustainable business practices that are environmentally responsible. On the one hand, the environmental performance of organizations should be enhanced through the implementation of effective control mechanisms utilizing Environmental Management Systems (EMS). On the other hand, there is a need to improve the exchange of information both internally and externally within the organization. The literature unequivocally presents evidence supporting the efficacy of an Emergency Medical Services (EMS) system. For example, Mitchell and Hill (2015) illustrate that EMS certified according to ISO 14001 have a positive influence on the voluntary reporting performance of South African companies. Nevertheless, Huang et al. (2024) found no evidence in their research to suggest that environmental management systems facilitate the adoption of environmentally sustainable practices by companies or enhance their reporting efforts. In light of the conflicting findings, an Environmental Management System (EMS) is not regarded as a limiting factor in the phenomenon of brownwashing, as it is purportedly utilized solely to cultivate an environmentally sustainable image.

2.1.9.3 CSR Committee

A CSR committee is a special focus group within a company that is predominantly concerned with the development, implementation, and monitoring of CSR aspects. The responsibility of the Corporate Social Responsibility (CSR) committee is to harmonize the social and sustainability concerns of stakeholders with the economic objectives of the corporation. The committee members serve as advisors to management, oversee business practices, and facilitate organizational change. A critical component of this monitoring process involves the examination of internal company data and the subsequent compilation of comprehensive reports. In particular, this data processing, as well as the monitoring by the CSR committee, may help to limit the principal–agent conflict that arises due to brownwashing. In Sect. 2.1.8 of the document, the establishment of a Corporate Social Responsibility (CSR) committee has been identified as a crucial factor in mitigating the practice of greenwashing.

The study conducted by Gull et al. (2023), which is deemed pivotal for the classification, fails to distinguish between greenwashing and brownwashing in its results. Although the distinction is initially articulated by Gull et al. (2023), the authors ultimately refer to the concept of decoupling. This suggests that a Corporate Social Responsibility (CSR) committee may also serve as a constraining factor in mitigating the practice of "brownwashing." Furthermore, the conclusions drawn by Gull et al. (2023) concerning the attributes of the Corporate Social Responsibility (CSR) committee are likewise accurate. As articulated by Gull et al. (2023), factors such as expertise, a greater workforce, autonomy, and the prolonged tenure of members contribute positively to mitigating decoupling. Huang et al. (2024) do not distinguish among these characteristics of the committee in their research. The findings of the study are, nonetheless, consistent with those of Gull et al. (2023), which indicate that the formation of a Corporate Social Responsibility (CSR) committee serves to mitigate the occurrence of brownwashing (Booth et al., 2024; Ghafoor & Gull, 2024; Gull et al., 2023).

2.1.10 CEO (Chief Executive Officer) Power

In the section addressing greenwashing within this study, prior reference has been made to the theory of managerial power and its associated implications for the phenomenon of greenwashing. According to the theory, managers leverage their authoritative positions to sway corporate decisions in a manner that benefits their interests. Considering the notion that brownwashing may evade the scrutiny of stakeholders, Chief Executive Officers would be afforded the opportunity to engage in opportunistic behavior concerning the organization. Conversely, the research conducted by Gull et al. (2023) presents evidence suggesting that the authority of the CEO serves to constrain the practice of brownwashing. This indicates that the authority of the CEO serves as a constraining element in the phenomenon of brownwashing.

2.1.11 Corporate Media Presence

A significant factor contributing to the hesitance of corporations to disclose their social and environmental commitments is associated with the preservation of the company's reputation. Owing to the cognitive limitations of stakeholders, the publications and statements of companies with a media presence are subjected to more rigorous scrutiny compared to those of companies that operate predominantly in the background and consequently do not garner media attention. This culminates in the conclusion that organizations engaging in active publication are increasingly under the scrutiny of stakeholders, thereby facing the potential risk of being perceived as hypocritical. Corporate communication is regarded as disingenuous when the statements disseminated by the organization do not align with its actual practices. In addition to the current actions undertaken, the company's prior conduct is also crucial for the credibility of its assertions. In relation to reputation, a differentiation is established between a general reputation and a context-specific reputation. The former emphasizes the overall reputation of the company. Conversely, when referring to a context-specific reputation, the organization is recognized for or due to a particular attribute or characteristic. If the company typically maintains a favorable reputation, the corporate image remains largely unblemished over an extended period, even in instances of uncertainty. Conversely, the reputation is specific to a particular context. Stakeholders anticipate a significant level of performance in the future within the domains for which the company is recognized. Should the company face scrutiny regarding this matter, it is likely to elicit adverse reactions from stakeholders and heightened media attention.

Carlos and Lewis (2017) conducted an investigation into the propensity of companies to publicly disclose their inclusion in the Dow Jones Sustainability Index

(DJSI). For companies listed in the DJSI, ecological and social criteria play a key role alongside economic factors. It emerged that companies with a context-specific reputation, in this instance, an environmental reputation, in particular, did not attract attention to their listing in the DJSI index. Carlos and Lewis (2017) contend that the proclamation will concentrate on enterprises that are already recognized for their environmentally sustainable business practices (Arranz et al., 2024; Chen & Guestrin, 2016). When a company first enters the media spotlight, its activities are subjected to rigorous scrutiny, and the organization faces the potential risk of having any hypocritical actions exposed. Nevertheless, these conclusions are, in certain respects, at odds with attribution theory. According to this theory, a company cannot be deemed hypocritical if its conduct remains consistent across various modalities and over an extended period. For instance, organizations that have historically engaged in comprehensive Corporate Social Responsibility (CSR) reporting will only be regarded as credible in the future if they persist in disclosing their activities. Similarly, organizations that consistently disclose their performance metrics are perceived as more credible than those that report only when deemed necessary. For a company to maintain its prominence in the future, it is imperative that it not be perceived as duplicitous.

Wei (2023) finds evidence that companies that publish a large number of environmental reports and have a reputation for their publications corresponding to their actual activities, or companies that are known for brownwashing, are the focus of investors. The assertion that a substantial volume of publications garners the attention of analysts or the media is similarly articulated in the research conducted by Carlos and Lewis (2017). Likewise, the statement that companies whose "walk and talk" concur are preferred by analysts is hardly startling. The assertion that analysts focus on companies that partake in brownwashing was initially articulated in the research conducted by Wei (2023). This arises from the tendency of "brownwashers" to present their social and environmental performance as being less significant than it truly is. In contrast to greenwashers, the information that has been disseminated has, in fact, been effectively integrated within the company. This indicates that, notwithstanding the potential underreporting of Corporate Social Responsibility (CSR) performance, the limited number of published activities has indeed been executed within the organization. If analysts subsequently employ brownwashers for their analyses, the accuracy of their forecasts is enhanced (Aronczyk et al., 2024; Du, 2015).

2.1.12 Board Network Centrality

Network theory examines the interconnections among individual members within a network and the corresponding benefits that arise from these relationships. The benefits stem from the dissemination of information within the network, which provides members with access to essential resources in the form of data. It is essential to acknowledge that there exist various relationships within the network, characterized by differing ranks and positions among its members. The greater the influence or power an individual possesses within a group, the more substantial the information content becomes, thereby enhancing the advantages derived from the network. Consequently, effective positioning is characterized as "network centrality." It is of significant significance for companies that the management board possesses an extensive network, which facilitates access to information both within and beyond the peer group. The literature indicates that a company with board members possessing network centrality is distinguished by enhanced advantages in both lending and merger and acquisition activities. However, enhanced financial performance is not the sole attribute. The research results indicate a positive correlation between board centrality and the social efficacy of a company. This indicates that "network centrality" provides a benefit from both financial and social standpoints.

Zhao et al. (2022) investigate the impact of network centrality on the Chinese capital market during the period from 2009 to 2018 (Zhao et al., 2022). They ascertain that, in addition to the networking among board members, the existing regulations pertaining to Corporate Social Responsibility (CSR) reporting are of paramount importance. This phenomenon can be attributed to the introduction of a guideline for voluntary reporting in China at the commencement of the observation period, specifically in 2009. The regulations, if they were indeed in place, were articulated in an ambiguous manner, and the authorities failed to conduct any verifications regarding the accuracy of the published materials. The diminished pressure indicated that numerous voluntary actions were predominantly symbolic in character. It was not until 2015 that the Chinese government enacted an environmental law that considerably strengthened the regulations pertaining to these issues.

The observations of Zhao et al. (2022) covered both the periods of less restrictive and strict CSR regulation. It became evident that under liberal corporate social responsibility regulations, the centrality of board networks exhibited a positive correlation with greenwashing; however, no association with brownwashing was observed. As regulatory measures intensified, the research findings indicate that network centrality exhibited a negative correlation with greenwashing and a positive correlation with brownwashing. This phenomenon can be elucidated by the introduction of stringent environmental regulations in 2015, which resulted in heightened scrutiny of corporate assurances and their practices related to corporate social responsibility (CSR) (Ruiz-Blanco et al., 2022; Trim & Jones, 2025; Zhao et al., 2022).

Zhao et al. (2022) argue that the controls reveal inequitable business practices concerning CSR, which diminish the company's authoritative standing within the network. The firm would thereafter forfeit its access to information exchange and, therefore, any associated advantages (Zhao et al., 2022). This suggests that during periods of stringent environmental laws, corporations may hesitate to disclose CSR activities to maintain their centrality within board networks. The study conducted by Carlos and Lewis (2015) corroborates these results. A corporation attracts attention via its publications, resulting in heightened scrutiny. Conversely, Zhao et al. (2022) observe that when controls are insufficient, corporations may issue reports of a symbolic character without restraint, since these publications are not subject to oversight. The authors contend that board network centrality amplifies the prevalence of greenwashing by heightening information asymmetry between corporations and stakeholders, thus complicating the detection of superficial CSR initiatives. Specifically, in nations like China, where the research was undertaken, boards of directors may use their political power to obstruct inquiries into their operations. The network facilitates the sharing of information among CEOs on opportunistic activities. The principal–agent conflict and a pivotal position within a network facilitate opportunistic behaviors, including greenwashing. To prevent redundancy, board network centrality was not included in the greenwashing part of the study.

2.1.13 Audit Remuneration

An audit is often conducted by an auditor and comprises a financial report that offers various stakeholders a comprehensive understanding of the company's economic condition. This insight need not pertain to a company's economic condition. An instance is the "Non-Financial Reporting Directive" (NFRD) instituted by the EU. This EU rule mandates that enterprises with over 500 workers must include non-financial information in their annual reports. These themes include environmental and social dimensions, adherence to human rights, and strategies to mitigate corruption and bribery. The primary objective of this initiative is to establish openness and accountability. The execution of such an audit incurs expenses and, as previously stated, is conducted by a reputable auditing company. Audit fees include the expenses associated with the report, mostly consisting of the projected work required for its preparation, a risk premium contingent upon the client's particulars, the auditor's reputation and image, and the prevailing economic conditions. Research conducted by the European Commission indicated that a medium-sized company incurs an average annual cost of EUR 40,000 for an NFRD audit. Due to these substantial costs, corporations endeavor to reduce the expenditures associated with an audit.

The fundamental concept of an audit is to diminish the information asymmetry between the organization and its stakeholders. The audit guarantees that all financial and off-balance sheet ties, including ESG efforts, are accurately and comprehensively reflected in the annual report. This provides stakeholders with profound insight into the organization, thereby reducing information asymmetry. The literature associates the phrase "accounting conservatism" with the substantial expenses associated with an audit. This phrase denotes the tendency of corporations and auditors to use a cautious methodology in asset presentation to mitigate agency expenses stemming from knowledge asymmetry (Fernandez, 2025; Gomes et al., 2024; Huang et al., 2024). Auditors prefer a cautious presentation since an excessive overvaluation of assets is more likely to result in legal challenges than an undervaluation. Conservative valuation methodologies are observable beyond only financial dimensions. Social pressure and several other causes have made the adoption of CSR activities in corporations standard. To issue a certified audit, it is essential to accurately consider both economic factors and social and environmental criteria. The audit procedure has been elevated, resulting in an increased workload for preparation and a heightened chance of misassessment. Thus, the costs required for an audit increase with its complexity and extent. The literature is polarized on these issues. Some research indicates that CSR programs elevate the work needed for report preparation and heighten the risk of auditing companies. Conversely, it is believed that non-monetary information, such as data about CSR commitment, enhances the assessment of business risk pricing (Liu & Li, 2024; Yu et al., 2020).

Adeneye et al. (2022) investigate this in their study and conclude that conservative accounting results in reduced audit fees. The premise that auditing firms exert less effort with prudent customers, resulting in reduced audit costs, is also evident in the context of organizations' CSR initiatives. Consequently, social and environmental performance enhances the intricacy of the audit process and exposes auditors to elevated risk. Auditing companies address the heightened audit risk associated with CSR by imposing elevated fees. According to the research conducted by Robertson et al. (2023) and the results of Adeneye et al. (2022), it can be said that auditors and corporations collaborate closely, possibly presenting their CSR initiatives in a conservative manner, hence resulting in a lack of transparency and the occurrence of greenwashing.

2.1.14 Characteristics of the Social or Environmental Certificate/Label

It was previously noted that firms can exploit stakeholder uncertainty about credentials, thereby engaging in greenwashing. This section of the thesis examines the attributes of the certificates and the consequent implications derived from the research of Carlos and Lewis (2017) and Delmas and Grant (2010) on brownwashing. The probability that a corporation would refrain from publicly disclosing an obtained certificate, thus participating in brownwashing, is associated, among other factors, with the credibility of the evidence. A certificate is deemed authentic if it is endorsed by the firm and its stakeholders, particularly the media, thereby achieving social recognition. Carlos and Lewis (2017) contend that firms often disseminate credentials that have not attained legitimacy status. This indicates that stakeholders are less familiar with the requirements imposed on the organization by the certifications and papers compared to those associated with valid certificates (Oguntona, 2024; Torelli et al., 2020; Xia et al., 2023; Xue et al., 2024). This implies that with less esteemed certifications, discrepancies between the stipulations outlined in the certificate and the company's real CSR actions are more prone to remain concealed. Conversely, this implies that authentic credentials are more susceptible to brownwashing, since stakeholders are aware of their particulars and may expose the company's hypocritical conduct. Delmas and Grant (2010) enhance their research by distinguishing between "eco-labeling" and eco-certification. The authors analyze the economic ramifications of the two words and explore the lack of honest communication about expensive eco-certifications in some instances. Although the research focuses on the wine sector, significant implications may be inferred about the broader topic of brownwashing.

Consumers of a product are inherently absent throughout the manufacturing process and hence cannot assert conformity with ecologically sustainable production standards. Eco-labels were used to mitigate information asymmetry. If a label is assigned by an autonomous institution based on clear and understandable criteria and then evaluated by a third party, the legitimacy of this designation is difficult to contest. A label established only by the manufacturer has little importance for genuine adherence to ecologically particular requirements. Moreover, the research indicates that many customers prioritize the perceived private benefits of ostensibly better quality items above the genuine environmental considerations when making purchasing decisions. Consumers will only accept a price premium for items with an "eco-label" if it offers a personal advantage. Certain corporations exploit the heightened propensity to purchase associated with a price premium by creating their own eco-labels. With the presence of both credible and opaque labels on the market, there exists a danger that consumers may get disoriented about the available environmental labels, leading to a potential loss of understanding of their significance. Delmas and Grant (2010) discovered that the price of wine decreased by 20% after marketing with an eco-label (Delmas & Burbano, 2011). It is important to acknowledge that customers' associations with "green wine" substantially influence the price reduction. Nonetheless, this indicates the lack of information and the ambiguity customers have with eco-labels. To prevent a comparable decline in prices, several companies refrain from affixing an environmental label to their goods, notwithstanding compliance with certain regulations. The attributes of a label are a crucial factor in brownwashing.

Unlike an "eco-label," an "eco-certification" is only granted by an accredited third party. The certified firm is prohibited from using the certification logo on its goods for marketing reasons. A prevalent criterion that differentiates sustainable management practices is certification by ISO 14001. Upon comparing eco-labels and certificates, it is evident that a label fundamentally functions to mitigate information asymmetry. The fundamental concept is that the customer is apprised of any sustainable manufacturing requirements throughout the purchase process. A certification that has been obtained is not immediately apparent to customers. ISO certification is used to identify and rectify inefficiencies within the organization, enhance product quality, and establish sustainable standards. This suggests that eco-certifications create a disparity between communication and actual practices about the company's environmental performance. The certification is not affixed directly to the goods like a label. Nonetheless, there are other methods to communicate the certification to stakeholders, including a post on the corporate website or inclusion in the sustainability report. Delmas and Grant (2010) discovered in their research that obtaining an environmental certification elevates the price of wine by around 13% on average. As a result, producers have no reason to conceal an environmental certificate acquired from a reputable certification body, such as ISO 14001 certification, from their stakeholders (Banihashemi et al., 2024; Dowell & Lyon, 2024; Prikshat et al., 2024).

These findings, in turn, contradict the research work of Carlos and Lewis (2017). According to this work, legitimate certificates, such as ISO certification, are particularly affected by brownwashing. In comparison to Delmas and Grant (2010), the authors argue that in the case of reputable certificates about which stakeholders are informed, inconsistencies between company practices and the actions described in the certificate could be discovered by interest groups. The authors' statements contradict each other, with each line of argument finding its justification.

2.1.15 Brownwashing as a Back-up for Reputation Crises

Legitimacy theory posits that firms consistently endeavor to uphold their reputation. Should the corporation commit any errors, there exists a danger that stakeholders or the public may use these errors to exert pressure on the organization. This pressure may manifest as a request for the organization to rectify its acts or assume accountability for them. To avert the looming reputation problem, the corporation would enhance its communication of good attributes to mitigate any allegations. The literature discusses the concept of "impression management" in this context. This pertains to the strategic endeavors of organizations that highlight the social and beneficial ideals of the enterprise via rhetorical or symbolic acts. Brownwashing may function as a supplementary strategy in this setting alongside obtaining the certifications. Acquired certifications that demonstrate the company's social or environmental commitment might be used to counteract pressure from stakeholders about these issues.

The certificate obtained from the corporations is not disclosed upon receipt due to their strategic actions. The certification is disclosed only when the firm is under external pressure. Due to the certificate, the besieged firm may, notwithstanding the allegations, cultivate the perception of being reliable and adhering to the mandated social or environmental norms. In this context, brownwashing may be seen as a strategic maneuver when favorable CSR accomplishments are intentionally concealed to mitigate potential reputational harm in the event of uncertainty.

2.1.16 Brownwashing to Hide Expenses

Milton Friedman (2007) asserts that a company's only responsibility is to maximize profits for its owners. Companies should not be compelled to fulfill social or civic responsibilities. Corporate social responsibility would thus hinder corporate efficiency, eventually jeopardizing competitiveness and the tenets of the free market. Friedman asserts that profit maximization is the only kind of corporate social responsibility. Friedman characterized other forms of CSR as "hypocritical window dressing." Years later, research indicates that the perspectives of several shareholders about CSR and corporate profit maximization have remained mostly unchanged. Jacobs et al. (2019) demonstrated that corporate voluntary emission reduction initiatives resulted in a statistically significant negative return of 0.95%. Fisher-Vanden and Thorburn (2011) presented another instance of shareholders' adverse response to voluntary CSR initiatives. Their analysis indicates that corporate statements on voluntary involvement in climate protection initiatives and greenhouse gas emission reduction resulted in a 1% decline in the relevant shares.

Implementing a CSR initiative undoubtedly incurs significant expenses. Shareholders may also incite concerns that CSR initiatives are being prioritized over other firm ideas, which would therefore be marginalized yet would directly enhance shareholder value. Literature suggests that the short-term return on investment in CSR initiatives is negative, since the adoption of environmentally sustainable technology is linked to substantial expenses. This fosters stability and sustainable development for all stakeholder groups. The enduring competitiveness fostered by CSR initiatives, contrasted with the adverse perceptions of some shareholders toward CSR, may entice corporations to resort to brownwashing (Chen et al., 2023; Kumar et al., 2023; Leboffe et al., 2023).

2.2 Consequences of Brownwashing

In the previous section of the book, detailed attention was paid to the determinants of brownwashing. Key influencing factors were identified that promote and limit the occurrence of brownwashing in companies. The focus below is on the effects that brownwashing can have on company performance.

2.2.1 Consumer Perception

Ellen et al. (2006) found in their study that a company's CSR performance usually stems from two different motives. On the one hand, stakeholder-oriented reasons play an important role. On the other hand, the company's value orientation. In the former, the company is committed to CSR in order to improve the organization's image and reputation externally. This motive, which refers in particular to legitimacy theory, is debated in detail in the chapter on greenwashing, which is why it is not discussed in more detail in this part of the thesis. For this section of the thesis, the value orientation of a company and, subsequently, the intrinsic motivation are essential. With reference to the corporate social responsibility of a company, the intrinsic motivation of an organization refers to those actions that the company does not take due to external constraints, but rather from the "internal" corporate values, motives, and beliefs. If an organization decides for certain reasons to present its CSR performance as less pronounced than it actually is in reality, according to the literature, it is taking a discreet position.

This discreet, or concealing, position implies that the CSR activity does not result in any obvious benefit for the company and that it has therefore arisen due to the company's intrinsic motivation. Ginder et al. (2021) find evidence that restrained (brownwashing) and transparent reporting have similarly positive effects on stakeholder trust, and thus, an indifferent purchase intention prevails among consumers. The prerequisite for this, however, is that both strategies are intrinsically motivated. This implies that the way in which reporting is done is secondary. The primary concern for stakeholders is the authenticity, consistency, and honesty of the information. Given the identical output achieved by the two strategies, the company should weigh up which form of reporting suits the company best. However, the research results also make it clear that particularly cautious CSR reporting can lead to consumers not perceiving any social and environmental performance on the part of the company at all and thus not perceiving any intrinsic motivation that leads to a purchase decision. Ginder et al. (2021) thus make it clear in their study that brownwashing does not negatively influence consumers' purchase decisions, as less pronounced communication of CSR performance leads consumers to conclude that companies are intrinsically motivated (Baah et al., 2023; El Koshiry et al., 2023; Eliwa et al., 2021; He et al., 2023; Karagiannis et al., 2022; Prabowo et al., 2017).

Upon reevaluating the argument presented by Testa et al. (2018), it is evident that by intentionally obscuring or withholding information, the corporation forfeits the many benefits associated with the public disclosure of CSR operations. These benefits are conferred to the organization only if stakeholders really recognize them. The organization engages in brownwashing by offering services while deliberately withholding communication about them, resulting in a lack of recognition by the public and leading to the company's classification as "brown." This renders the firm unappealing, especially to environmentally conscious consumer segments, perhaps resulting in purchasing decisions favoring competitors with a more favorable reputation. The established trend of "green finance" in the market, which involves targeted investments in sustainable assets, should not be overlooked. Companies engaging in brownwashing remove themselves from this asset class and hence forfeit the advantages of the trend.

2.2.2 Implications for Corporate Performance

Companies that have integrated sustainability into their company principles and communicate this honestly and openly may capitalize on investment trends favoring green and sustainable enterprises and initiatives. Appealing financing prospects emerge from the "green finance" movement within the capital market and from governmental incentives and funding initiatives. The latter seeks to assist socially and ecologically responsible enterprises in their operations to facilitate the transition to a sustainable society. Government incentives and financing schemes may take the shape of grants, tax perks, or other financial advantages.

From this perspective, it might be said that corporations involved in brownwashing are ostensibly excluded from these financial resources. The research indicates that the shift toward green and sustainable investments is also motivated by the perceived reduced risk of default and reputational concerns. Default risk indicates the possibility of an investor losing their invested capital, while reputation risk signifies the possible adverse effects on the investor's reputation or status related to the asset in question. Eliwa et al. (2021) and Goss and Roberts (2011) contend in their research on default risk that the information asymmetry resulting from brownwashing complicates shareholders' ability to accurately assess business risk. Companies without evidence of social or environmental stewardship, or those potentially contributing to environmental degradation, often encounter an elevated risk premium. This enables the investor to safeguard against risks arising from insufficient social and environmental responsibility. The outcome is increased difficulty in capital acquisition, which escalates corporate expenditures and eventually jeopardizes the development and competitiveness of the firm involved in greenwashing (Eliwa et al., 2021).

The literature is polarized over the matter of capital acquisition by lending institutions. Erragragui (2018) identifies data indicating that corporate governance adversely affects the cost of capital. Business governance pertains to business practices that ensure adherence to morally and ethically acceptable norms. Systems and procedures designed to uphold ethical and moral standards often lead to elevated expenses. Enhanced transparency mandates, extensive reporting obligations, and related compliance rules serve as cost drivers in this setting, among other factors. Organizations exhibiting superior corporate governance, therefore, incur elevated capital expenses. Moreover, disputes over corporate governance do not substantially affect the cost of capital. Consequently, investors place less emphasis on effective corporate governance than is often assumed. The research by Eliwa et al. (2021) likewise corroborates these results. The analysis indicates that there is no substantial negative correlation between the advantages of corporate governance and the cost of debt. This suggests that the disclosure of corporate governance operations may not enhance capital acquisition and may potentially impose an extra barrier. To enhance business performance, corporations may choose to conceal expenditures and thus engage in brownwashing (Blome et al., 2017; Jiang et al., 2024; Nyilasy et al., 2014; Tong et al., 2024).

Nonetheless, concerning socially and ecologically responsible business practices, Eliwa et al. (2021) identified a markedly negative correlation with a company's cost of capital. Specifically, environmentally sustainable business practices substantially contribute to a decrease. This indicates that, alongside economic factors, lending institutions consider adherence to social and environmental norms in capital allocation, resulting in "brown companies" facing a financial disadvantage regarding capital expenses. The incorporation of CSR criteria in the loan process is sometimes driven by heightened societal interest in sustainable corporate practices and the looming reputational damage associated with non-compliance. Historically, the dominant perspective was that credit institutions alone evaluated verifiable and objective criteria, such as profitability, debt, and liquidity, when approving loans; however, this viewpoint has been augmented by the consideration of reputation. As a result, financial institutions may be seen as endorsers of ecologically detrimental activities if they provide loans to enterprises with poor CSR performance. Companies that do not align with economic lending standards but adhere to social and ecological norms may face loan denial. Consequently, it may be said that brownwashing has inherently detrimental effects on capital acquisition.

2.2.3 Analyst Projections

Analysts need a substantial volume of reliable data to provide an accurate evaluation of a firm. Alongside the economic data, social and economic factors also have substantial significance. Concerning brownwashing, there exists a possibility that firms may obscure their true operations despite the availability of substantial CSR services. The prudent strategy in the release results in little information being accessible to analysts, leading to significant information asymmetry between corporations and analysts. The disproportionate information ratio may result in a degradation of the accuracy of analyst projections. Research indicates that analysts mitigate information asymmetry via their specialized and sector expertise, resulting in analyst predictions maintaining a degree of accuracy even in instances of greenwashing.

2.2.4 Implications for Corporate Valuation

Hawn and Ioannou (2016) investigate the correlation between "talk and walk" and its effects on a company's market value in their research (Hawn & Ioannou, 2016). The authors, aligning with conventional perspectives, assert that the implementation and effective communication of CSR activities positively impact a company's market value; nonetheless, the study's conclusions are especially pertinent about the unfavorable disparity between rhetoric and action. Hawn and Ioannou (2016) demonstrate that brownwashing results in a decline in market value. The analysis indicates that the most significant disparity between "walk" and "talk" results in a market value that is 26.5% lower than the average market value of the other firms. A firm that does not report a discrepancy has a market worth that is 40% more than that of the company most significantly engaged in greenwashing. The research indicates that over 50% of the analyzed firms participated in brownwashing (Budzanowska-Drzewiecka & Proszowska, 2024; Ruiz-Blanco et al., 2022; Yu et al., 2020; Zhao et al., 2022).

Consistent with Hawn and Ioannou (2016), Ginder et al. (2021) assert that inadequately articulated CSR activities may fail to be recognized and valued by stakeholders. Owing to the principal–agent dilemma and the ensuing information asymmetry, external interest groups cannot discern a successful "walk" without accompanying "talk," implying that, at worst, stakeholders fail to recognize any meaningful, much less symbolic, CSR initiatives. Hawn and Ioannou (2016) found that stakeholders anticipate a certain degree of CSR engagement, resulting in brownwashing being linked to regulatory concerns, diminished legitimacy, unhappiness, and unmet expectations. This, therefore, is shown in the markedly diminished valuation of the firm. Nonetheless, the conclusions of Hawn and Ioannou (2016) may only be deemed revolutionary in sectors that are attuned to social and environmental performance, such as the extractive industries and mining.

References

Adeneye, Y. B., Kammoun, I., Wahab, S. N. A. A. (2022). Capital structure and speed of adjustment: the impact of environmental, social and governance (ESG) performance. *Sustainability Accounting, Management and Policy Journal, 14*(5): 945–977. https://doi.org/10.1108/SAMPJ-01-2022-0060

Aronczyk, M., McCurdy, P., & Russill, C. (2024). Greenwashing, net-zero, and the oil sands in Canada: The case of pathways Alliance. *Energy Research and Social Science, 112*. https://doi.org/10.1016/J.ERSS.2024.103502

Arranz, C. F. A., Arroyabe Marta, F., & Fernandez de Arroyabe, J. C. (2024). Organisational transformation toward circular economy in SMEs. The effect of internal barriers. *Journal of Cleaner Production, 456*, 142307. https://doi.org/10.1016/J.JCLEPRO.2024.142307

Baah, C., Agyabeng-Mensah, Y., Afum, E., & Kumi, C. A. (2023). Do circular economy practices accelerate CSR participation of SMEs in a stakeholder-pressured era? A network theory perspective. *Journal of Cleaner Production, 394*. https://doi.org/10.1016/J.JCLEPRO.2023.136348

Banihashemi, S., Meskin, S., Sheikhkhoshkar, M., et al. (2024). Circular economy in construction: The digital transformation perspective. *Cleaner Engineering and Technology, 18*. https://doi.org/10.1016/j.clet.2023.100715

Bernard, S., & Nicolau, J. L. (2022). Environmental certification and hotel market value. *International Journal of Hospitality Management, 101*. https://doi.org/10.1016/j.ijhm.2021.103129

Blome, C., Foerstl, K., & Schleper, M. C. (2017). Antecedents of green supplier championing and greenwashing: An empirical study on leadership and ethical incentives. *Journal of Cleaner Production, 152*, 339–350. https://doi.org/10.1016/J.JCLEPRO.2017.03.052

Booth, H., Milner-Gulland, E. J., McCormick, N., & Starkey, M. (2024). Operationalizing transformative change for business in the context of nature positive. *One Earth, 7*, 1235–1249. https://doi.org/10.1016/J.ONEEAR.2024.06.003

Brammer, S., & Pavelin, S. (2006). Corporate reputation and social performance: The importance of fit. *Journal of Management Studies, 43*, 435–455. https://doi.org/10.1111/j.1467-6486.2006.00597.x

Budzanowska-Drzewiecka, M., & Proszowska, A. (2024). Greenwashing. In *Reference module in social sciences*. Elsevier. https://doi.org/10.1016/B978-0-443-13701-3.00146-8

Carlos, W. C., & Lewis, B. W. (2017). Strategic silence: Withholding certification status as a hypocrisy avoidance tactic. *Administrative Science Quarterly, 63*(1), 130–169. https://doi.org/10.1177/0001839217695089. (Original work published 2018).

Chen, P., & Dagestani, A. A. (2023). Greenwashing behavior and firm value – From the perspective of board characteristics. *Corporate Social Responsibility and Environmental Management, 30*, 2330–2343. https://doi.org/10.1002/CSR.2488

Chen, T., & Guestrin, C. (2016). *XGBoost: A scalable tree boosting system*. ACM Digital Library. https://doi.org/10.1145/2939672.2939785

Chen, C., Okubo, R., Okawa, S., et al. (2023). The prevalence and risk factors of suicidal ideation in women with and without postpartum depression. *Journal of Affective Disorders, 340*, 427–434. https://doi.org/10.1016/j.jad.2023.08.051

Delmas, M. A., & Grant, L. E. (2010). Eco-Labeling Strategies and Price-Premium: The Wine Industry Puzzle. *Business & Society, 53*(1), 6–44. https://doi.org/10.1177/0007650310362254

Delmas, M. A., & Burbano, V. C. (2011). The drivers of greenwashing. *California Management Review, 54*, 64–87. https://doi.org/10.1525/CMR.2011.54.1.64

Dowell, G., & Lyon, T. (2024). Beliefs matter: Local climate concerns and industrial greenhouse gas emissions in the United States. *Journal of Business Ethics*. https://doi.org/10.1007/S10551-024-05619-W

Du, X. (2015). How the market values greenwashing? Evidence from China. *Journal of Business Ethics, 128*, 547–574. https://doi.org/10.1007/S10551-014-2122-Y

El Koshiry, A., Eliwa, E., Abd El-Hafeez, T., & Shams, M. Y. (2023). Unlocking the power of blockchain in education: An overview of innovations and outcomes. *Blockchain: Research and Applications, 4*. https://doi.org/10.1016/j.bcra.2023.100165

Ellen, P., Webb, D., & Mohr, L. (2006). Building corporate associations: Consumer attributions for corporate socially responsible programs. *Journal of the Academy of Marketing Science, 34*, 147–157. https://doi.org/10.1177/0092070305284976

Eliwa, Y., Aboud, A., & Saleh, A. (2021). ESG practices and the cost of debt: Evidence from EU countries. *Critical Perspectives on Accounting, 79*, 102097. https://doi.org/10.1016/J.CPA.2019.102097

Erragragui, E. (2018). Do creditors price firms' environmental, social and governance risks? *Research in International Business and Finance*, Elsevier, *45*(C), 197–207.

Fernandez, V. (2025). Corporate greenwashing and green management indicators. *Environmental and Sustainability Indicators, 100599*. https://doi.org/10.1016/J.INDIC.2025.100599

Fisher-Vanden, K., & Thorburn, K. S. (2011). Voluntary corporate environmental initiatives and shareholder wealth. *Journal of Environmental Economics and Management*, Elsevier, *62*(3), 430–445.

Friedman, M. (2007). *The social responsibility of business is to increase its profits*. https://doi.org/10.1007/978-3-540-70818-6_14

Gacek, J. (2020). Corporate greenwashing and Canada goose: Exploring the legitimacy–aesthetic nexus. *International Journal for Crime, Justice and Social Democracy, 9*. https://doi.org/10.5204/IJCJSD.V9I2.1385

Ghafoor, A., & Gull, A. A. (2024). Do co-opted boards protect CEOs from ESG controversies? *Finance Research Letters, 63*. https://doi.org/10.1016/j.frl.2024.105263

Ginder, W., Kwon, W.-S., & Byun, S.-E. (2021, March). Effects of internal–external congruence-based CSR positioning: An attribution theory approach. *Journal of Business Ethics*, Springer, *169*(2), 355–369.

Gomes, M., Marsat, S., Peillex, J., & Pijourlet, G. (2024). Does religiosity influence corporate greenwashing behavior? *Journal of Cleaner Production, 434*. https://doi.org/10.1016/J.JCLEPRO.2023.140151

Goss, A., & Roberts, G. S. (2011, July). The impact of corporate social responsibility on the cost of bank loans. *Journal of Banking & Finance*, Elsevier, *35*(7), 1794–1810.

Gull, A. A., Hussain, N., Khan, S. A., et al. (2023). The power of the CEO and environmental decoupling. *Business Strategy and the Environment, 32*, 3951–3964. https://doi.org/10.1002/BSE.3347

Hawn, O., & Ioannou, I. (2016). Mind the gap: The interplay between external and internal actions in the case of corporate social responsibility. *Strategic Management Journal, 37*, 2569–2588. https://doi.org/10.1002/SMJ.2464

He, F., Liu, G., Hao, J., & Li, Y. (2023). CSR performance and firm idiosyncratic risk in a data-rich environment: The role of retail investor attention. *Journal of International Financial Markets, Institutions and Money*, 101877. https://doi.org/10.1016/J.INTFIN.2023.101877

Huang, C.-L., & Kung, F.-H. (2010, March). Drivers of environmental disclosure and stakeholder expectation: Evidence from Taiwan. *Journal of Business Ethics, 96*, 435–451. https://doi.org/10.1007/s10551-010-0476-3

Huang, Z., Shi, Y., & Jia, M. (2025). Greenwashing: A systematic literature review. *Accounting and Finance*, Accounting and Finance Association of Australia and New Zealand, *65*(1), 857–882.

Huang, Y., Xiong, N., & Liu, C. K. (2024). Renewable energy technology innovation and ESG greenwashing: Evidence from supervised machine learning methods using patent text. *Journal of Environmental Management, 370*. https://doi.org/10.1016/j.jenvman.2024.122833

Hussain, T., Wang, D., & Benqian, L. (2023). Examining the role of responsible management, CSR, and TQM in enhancing renewable energy projects: An empirical analysis. *Acta Ecologica Sinica*. https://doi.org/10.1016/j.chnaes.2023.06.010

Jacobs, B. L., & Finney, B. (2019). Defining sustainable business—beyond greenwashing. *Virginia Environmental Law Journal, 37*(2), 89–131. https://www.jstor.org/stable/26742666

Jiang, C., Li, X., Xu, Q., & Liu, J. (2024). Does environmental protection tax impact corporate ESG greenwashing? A quasi-natural experiment in China. *Economic Analysis Policy, 84*, 774–786. https://doi.org/10.1016/J.EAP.2024.09.029

Karagiannis, I., Vouros, P., Sioutas, N., & Evangelinos, K. (2022). Mapping the maritime CSR agenda: A cross-sectoral materiality analysis of sustainability reporting. *Journal of Cleaner Production, 338*. https://doi.org/10.1016/j.jclepro.2021.130139

Katsaros, K., Marggraf, C., Ebi, K. L., et al. (2024). Exploring interconnections: A comprehensive multi-country analysis of climate change, energy demand, long-term care, and health of older adults. *Maturitas, 184*. https://doi.org/10.1016/j.maturitas.2024.107961

Kim, E. H., & Lyon, T. P. (2015). Greenwash vs. Brownwash: Exaggeration and undue modesty in corporate sustainability disclosure. *Organization Science, 26*, 705–723. https://doi.org/10.1287/ORSC.2014.0949

Kumar, N. R., Arias, M. P., Leitner, K., et al. (2023). Assessing the impact of telehealth implementation on postpartum outcomes for black birthing people. *American Journal of Obstetrics & Gynecology MFM, 5*. https://doi.org/10.1016/j.ajogmf.2022.100831

Lagoarde-Ségot, T. (2024). Greenwashing and sustainable finance: An approach anchored in the philosophy of science. *Current Opinion in Environment Sustainability, 66*. https://doi.org/10.1016/J.COSUST.2023.101397

Leboffe, E. N., Pietragallo, H. C., Liu, G., et al. (2023). The impact of the 2015 ACOG screening guidelines on the diagnosis of postpartum depression among privately insured women. *Journal of Affective Disorders, 328*, 103–107. https://doi.org/10.1016/j.jad.2023.02.020

Li, W., Shi, C., Xiao, Z., & Zhang, X. (2024). Bridging the green gap: How digital financial inclusion affects corporate ESG Greenwashing. *Finance Research Letters, 106018*. https://doi.org/10.1016/J.FRL.2024.106018

Liu, Z., & Li, X. (2024). The impact of bank fintech on ESG greenwashing. *Finance Research Letters, 62*. https://doi.org/10.1016/j.frl.2024.105199

Mandal, R., & Murthy, A. (2021). CSR in the post pandemic era: The dual promise of ESG investment and investor stewardship. *Indian Law Review, 5*, 229–249. https://doi.org/10.1080/24730580.2021.1899627

Marquis, C., Toffel, M., & Zhou, Y. (2016). Scrutiny, norms, and selective disclosure: A global study of greenwashing. *Organization Science, 27*, 483–504. https://doi.org/10.1287/orsc.2015.1039

Miao, G., Chen, G., Wang, F. & Das, A. K. (2023). The Effect of Corporate Greenwashing on Employees' Environmental Performance: Person–Organization Values Fit Perspective. *Sustainability, 15*(4), 3498. https://doi.org/10.3390/su15043498

Mitchell, R., & Boyle, B. (2015). Professional diversity, identity salience and team innovation: The moderating role of open-mindedness norms: Professional diversity and team innovation. *Journal of Organizational Behavior, 36*. https://doi.org/10.1002/job.2009

Montgomery, A., Lyon, T., & Barg, J. (2023b). No end in sight? A greenwash review and research agenda. *Organization & Environment., 37*, 108602662311689. https://doi.org/10.1177/10860266231168905

Negri, M., Cagno, E., Colicchia, C., & Sarkis, J. (2021). Integrating sustainability and resilience in the supply chain: A systematic literature review and a research agenda. *Business Strategy and the Environment, 30*, 2858–2886. https://doi.org/10.1002/bse.2776

Nyilasy, G., Gangadharbatla, H., & Paladino, A. (2014). Perceived greenwashing: The interactive effects of green advertising and corporate environmental performance on consumer reactions. *Journal of Business Ethics, 125*, 693–707. https://doi.org/10.1007/S10551-013-1944-3

Oguntona, O. A. (2024). Evaluation of global research on greenwashing by scientometric indicators. *Sustainability and Toxicity of Building Materials*, 59–79. https://doi.org/10.1016/B978-0-323-98336-5.00004-2

Petrenko, O. V., Aime, F., Ridge, J., & Hill, A. (2016). Corporate social responsibility or CEO narcissism? CSR motivations and organizational performance. *Strategic Management Journal, 37*, 262–279. https://doi.org/10.1002/SMJ.2348

Prabowo, M. A., Iswaningtyas, A., Syofyan, E., et al. (2017). Board of directors and CSR disclosure in Indonesian banking industry: Does education matter. *International Journal of Trade and Global Markets, 10*, 322–338. https://doi.org/10.1504/IJTGM.2017.090280

Prikshat, V., Patel, P., Kumar, S., et al. (2024). Role of socio-cultural capital and country-level affluence in ethical consumerism. *Journal of Business Ethics*. https://doi.org/10.1007/S10551-024-05662-7

Qiao, D., Jiao, J., Khalid, N., & Ali, M. H. (2025). Supply chain concentration and corporate green innovation: Evidence from China. *Innovation and Green Development, 4*, 100202. https://doi.org/10.1016/J.IGD.2024.100202

Robertson, J. L., Montgomery, A. W., & Özbilir, T. (2023). Employees' Response to Corporate Greenwashing. *Business Strategy and the Environment, 32*(7), 4015–4027. https://doi.org/10.1002/bse.3351

Ruiz-Blanco, S., Romero, S., & Fernandez-Feijoo, B. (2022). Green, blue or black, but washing–what company characteristics determine greenwashing? *Environment, Development and Sustainability, 24*, 4024–4045. https://doi.org/10.1007/S10668-021-01602-X/TABLES/8

Sánchez, I., Hussain, N., Aibar-Guzmán, C., & Aibar-Guzmán, B. (2021). Assurance of corporate social responsibility reports: Does it reduce decoupling practices? *Business Ethics, the Environment & Responsibility, 31*, 118–138. https://doi.org/10.1111/beer.12394

Strand, R., & Freeman, R. E. (2015). Scandinavian cooperative advantage: The theory and practice of stakeholder engagement in Scandinavia. *Journal of Business Ethics, 127*, 65–85. https://doi.org/10.1007/S10551-013-1792-1

Testa, F., Boiral, O., & Iraldo, F. (2018). Internalization of environmental practices and institutional complexity: Can stakeholders pressures encourage greenwashing? *Journal of Business Ethics, 147*, 287–307. https://doi.org/10.1007/S10551-015-2960-2

Tong, Y., Lau, Y. W., & Binti Ngalim, S. M. (2024). Do pilot zones for green finance reform and innovation avoid ESG greenwashing? Evidence from China. *Heliyon, 10*. https://doi.org/10.1016/j.heliyon.2024.e33710

Torelli, R., Balluchi, F., & Lazzini, A. (2020). Greenwashing and environmental communication: Effects on stakeholders' perceptions. *Business Strategy and the Environment, 29*, 407–421. https://doi.org/10.1002/BSE.2373

Trim, I., & Jones, A. (2025). Do the world's largest companies integrate natural capital risks and opportunities into their operations in a meaningful way? *Biological Conservation, 302*. https://doi.org/10.1016/j.biocon.2024.110916

Wang, H., & Shen, H. (2024). Self-restraint or external supervision: Green bond issuance and greenwashing. *Research in International Business and Finance, 70*. https://doi.org/10.1016/J.RIBAF.2024.102402

Wang, Z., Hsieh, T. S., & Sarkis, J. (2018). CSR performance and the readability of CSR reports: Too good to be true? *Corporate Social Responsibility and Environmental Management, 25*, 66–79. https://doi.org/10.1002/CSR.1440

Wang, Y., Xing, C., & Zhang, L. (2024). Is greenwashing beneficial for corporate access to financing? Evidence from China. *Finance Research Letters, 65*. https://doi.org/10.1016/J.FRL.2024.105538

Wei, J. (2023). Does the "Greenwashing" and "Brownwashing" of corporate environmental information affect the analyst forecast accuracy? *Sustainability, 15*(14), 11461. https://doi.org/10.3390/su151411461

Xia, F., Chen, J., Yang, X., et al. (2023). Financial constraints and corporate greenwashing strategies in China. *Corporate Social Responsibility and Environmental Management, 30*, 1770–1781. https://doi.org/10.1002/CSR.2453

Xue, K., Chen, X., Sun, P., & Li, Y. (2024). Supply chain network centrality and corporate greenwashing behavior. *Finance Research Letters, 70*. https://doi.org/10.1016/j.frl.2024.106357

Ye, Y., Gao, M., & Luo, R. (2024). Bright is the moon over my home village: Chinese local executives and environmental decoupling. *Global Finance Journal, 62*. https://doi.org/10.1016/j.gfj.2024.101027

Yu, E. P., Van Luu, B., & Chen, C. H. (2020). Greenwashing in environmental, social and governance disclosures. *Research in International Business and Finance, 52*, 101192. https://doi.org/10.1016/J.RIBAF.2020.101192

Yu, X., Xiao, K., & Xu, T. (2023). Does ESG profile depicted in CSR reports affect stock returns? Evidence from China. *Physica A: Statistical Mechanics and its Applications, 627*. https://doi.org/10.1016/j.physa.2023.129118

Zhang, D. (2022). Green financial system regulation shock and greenwashing behaviors: Evidence from Chinese firms. *Energy Economics, 111*, 106064. https://doi.org/10.1016/J.ENECO.2022.106064

Zhang, M., & He, W. (2025). Encouraging or inhibiting: Can analyst attention reduce corporate greenwashing behavior? *Economic Analysis Policy, 85*, 943–962. https://doi.org/10.1016/J.EAP.2025.01.007

Zhang, J., Xu, D., & Chen, L. (2024). Enhancing supportive intention through perceived legitimacy: Social media influencer leadership and charismatic legitimization in CSR communication. *Public Relations Review, 50*. https://doi.org/10.1016/j.pubrev.2024.102511

Zhang, W., Yue, L., Qi, C., et al. (2024). Effects of coupling of groove microstructure and mixed wettability on flow boiling heat transfer enhancement of R134a in microchannels. *International Communications in Heat and Mass Transfer, 155*. https://doi.org/10.1016/j.icheatmasstransfer.2024.107571

Zhao, W., Zhong, M., Liao, X., et al. (2022). Board network and CSR decoupling: Evidence from China. *Frontiers in Psychology, 13*, 815341. https://doi.org/10.3389/FPSYG.2022.815341/BIBTEX

Zhen Huang & Yanxuan Shi & Ming Jia, 2025. Greenwashing: A systematic literature review, Accounting and Finance, Accounting and Finance Association of Australia and New Zealand, vol. 65(1), pages 857-882, March.

3 Board Characteristics and Greenwashing: An Integrated Financial Investigation Using Entropy Weight and TOPSIS Multicriteria Decision-Making

3.1 Greenwashing and Corporate Governance: A Dual Framework That Raises Controversy

Greenwashing is the practice of misleadingly suggesting to internal or external stakeholders that an organization engages in more ecologically sustainable activities than it really does. It is often seen in two essential areas: disclosures and marketing. The former relates to many types of reports, including sustainability reporting, investor communications, and climate-related disclosures, including those produced by publicly listed corporations or entities within the financial services and asset management industries (Lagoarde-Ségot, 2024; Li, 2024). The latter relates to the portrayal of businesses, including public relations, brand image, and advertising in the sectors of fashion, transportation, and consumer goods. Publicly traded firms, private enterprises, and non-profit organizations are all apprehensive about greenwashing, since it may appear in many communication forms and across several sectors. Besides deliberate greenwashing, it may also transpire inadvertently, like when management inadequately discloses sustainability matters that are not financially important (Peng & Xie, 2024).

This misinformation, regardless of whether it was done on purpose or not, causes confusion for investors, regulators, and customers. It may be difficult to tell the difference between what is considered authentic reporting of an organization's legitimate attempts to solve environmental, social, and governance (ESG) concerns and what is considered greenwashing. Environmental, social, and governance (ESG) issues exist in a rapidly changing environment that is filled with a range of players, each of whom has a different degree of competence and area of interest. The distinction becomes even more muddled when organizations that only concentrate on one environmental problem are accused of greenwashing. These groups may be experiencing a certain level of "tunnel vision." Defining and listing the many types of greenwashing, without being comprehensive, might help clarify the limits

K. Ragazou et al., *Transparency and Corporate Washing in the Age of Sustainability*, Management for Professionals,
https://doi.org/10.1007/978-3-031-96821-1_3

(Koemtzopoulos et al., 2025; Zournatzidou, 2024; Zournatzidou, Sklavos, Ragazou, & Sariannidis, 2024).

In Europe, especially in the banking industry, greenwashing has become much more common in recent years. In this area, greenwashing has been much more common over the last 5 years, with the number of incidents increasing threefold. As a result, it is understandable that regulators, national competition, and market authorities have acknowledged this. 359 publicly traded financial companies were chosen by the authors of this research to be rated for greenwashing. The banking sector in Southern Europe may be considered a "greenwasher" in the market. The existing sanctions structure in this nation, which is based on the one in Italy, may be seen as a kind of greenwashing. This may be due to the fact that the government does not have a dedicated regulatory framework that specifically handles greenwashing in the financial industry. Italy has a significant compliance gap, which is characterized by a lack of supervisory control for claims related to sustainability. Many financial laws and regulations, including those that deal with consumer protection and the prohibition of misleading advertising, indirectly address the issue of greenwashing in the financial sector (Gomes et al., 2024; Li et al. 2024; Zhao et al., 2024). There is no plan to establish a cohesive and well-structured framework for the financial industry. Instead, a number of European Union (EU) nations are implementing regulations against dishonest corporate practices under consumer law and broad sectoral principles in order to prevent greenwashing. Even yet, there is no clear advice on the best way to fight against greenwashing.

Greenwashing is not limited to Italy and southern Europe; it happens in other regions as well. The financial industry in the United Kingdom must also deal with the same problem. However, the government of the United Kingdom has recently put in place more extensive measures to combat greenwashing (Koemtzopoulos et al., 2025). This law will apply to all regulated businesses that make any kind of sustainability claim about their goods or services in any contact with consumers. The term "customer" includes both retail and business clientele. Financial goods and service providers must make statements that are "consistent with the sustainability profile of the product or service, equitable, transparent, and not misleading" in order to prevent greenwashing. The Environmental, Social, and Governance sourcebook (ESG) from the Financial Conduct Authority (FCA) includes the new legislation in ESG 4.3.1.

Furthermore, the financial sector in the Netherlands is among the least greenwashed in Europe. The Dutch financial companies are very certain that the market no longer allows for "greenwashing." This result is the consequence of the introduction of strict rules, especially during the most recent fiscal year. The Code for Sustainability Advertising (CSA) is a guideline that applies to all sustainability claims made in the financial sector, including those related to the environment and ethics. The word "sustainable" includes environmental issues in addition to labor conditions. The CSA is responsible for overseeing all parts of the product and service life cycle in the sector, including the management of garbage. Additionally, Corporate Digital Responsibility (CDR) is relevant to marketing strategies that include warnings about the environmental impact of certain financial goods or services (Liu & Li, 2024; Xue et al., 2024; Yin & Yang, 2024).

3.1.1 Main Features of Greenwashing

A recent academic study has found a number of characteristics that lead to greenwashing. Organizational factors are the features of a firm, such as its size, profitability, industry, market-to-book ratio, and leverage. People often assume that a company's size prevents it from engaging in greenwashing. It is more likely that non-governmental groups, activists, and the media will pressure major firms to adopt environmentally friendly practices, regardless of whether they are publicly or privately owned. This is because the problem is now more important to all parties involved, especially the customers. It is expected that these firms would have more visibility and would be required to follow stricter disclosure laws. Furthermore, big companies will be less inclined to participate in greenwashing since they are more vulnerable to being caught, suffering damage to their brand, and lacking credibility. They will also be more willing to provide more information about their environmental projects. The openness of environmental reporting is increased, and the effect of greenwashing is reduced by its size.

Some experts argue that there is a positive relationship between the distribution of thorough and trustworthy environmental information and the profitability of businesses. At the same time, there is conflicting information about profitability and greenwashing. Some scholars claim that companies that are more financially successful are more likely to engage in greenwashing because they have the financial capabilities to reduce reputational harm, for example, by settling legal bills or fines. There are two main aspects that contribute to the improvement of the market value of businesses by making environmental reporting more accurate and easier to understand: Investors showed more confidence in their decision-making and a decreased sense of information asymmetry. Furthermore, the cost of capital is reduced due to the lower amount of money spent on acquiring information. Even though this is true, the public's impression of corporate greenwashing is affected by the goals that environmental programs claim to have. When an ecologically friendly endeavor is motivated by economic incentives, skeptics are less likely to be critical of greenwashing.

Businesses that have a significant amount of debt in their capital structure are more likely to provide extra information on their financial health and environmental performance. This might suggest that there is a connection between greenwashing and the use of debt to hide one's financial situation. More and more creditors are trying to get complete information on the people who owe them money. As a result, companies that have a higher level of debt may have fewer financial resources available to do environmental reporting and study their ecological impact. This suggests that the danger of greenwashing may be higher due to the greater amount of power. To sum up, the industry in which a firm works has an impact on the process of greenwashing. This is particularly important for businesses that are sensitive to environmental issues, such as utilities and mining. These companies often struggle to achieve sustainability, are subject to stricter restrictions, and face higher brand damage. It is essential for bigger companies to implement improved reporting efforts and to respond to the growing demand from society. We believe that firms that have the potential to have a greater impact on the environment are less likely to participate in greenwashing.

3.2 Materials and Methods

3.2.1 Data

This research aims to investigate the impact of a novel set of governance factors that can affect the greenwashing score of 359 financial institutions in Europe (Table 3.1). In fact, we have implemented the entropy weight and Technique for Order of Preference by Similarity to Ideal Solution (TOPSIS) methods to examine this relationship. Data were retrieved from the database of Refinitiv Eikon, powered by Thomson Reuters, for the Fiscal Year 2024.

3.2.2 Entropy Weight Method

The TOPSIS technique is a systematic approach that is used to make decisions based on several criteria within a defined framework. Entropy weighting, the

Table 3.1 Criteria considered in TOPSIS analysis

Criteria	Description
C1. Board size	The total number of board members at the conclusion of the fiscal year
C2. Governance pillar score	The corporate governance pillar evaluates a company's procedures and processes that guarantee its board members and executives work in the best interests of long-term shareholders. It demonstrates a company's ability, via the implementation of optimal management practices, to govern and oversee its rights and obligations via the establishment of incentives and checks and balances to produce long-term shareholder value
C3. External consultants	Indicates the capacity of the board or its committees to engage external advisors or consultants independently of the management's consent
C4. Board conflicts of interest	Conflicts of interest among board members constitute a substantial ethical issue, arising when personal interests collide with the need to prioritize the organization's best interests. Evaluate if the board manages conflicts of interest or has a policy to resolve such conflicts at the board level
C5. Board effectiveness review	A board effectiveness review is a process that involves assessing and analyzing the effectiveness and efficiency of a company's board of directors in fulfilling its responsibilities and achieving corporate objectives. The objective of the evaluation is to confirm the effectiveness of the entire board. The evaluation should exclusively focus on the board's efficacy. The corporation is required to establish a precise timeline for this procedure
C6. Board oversight of health and safety	Board supervision of health and safety indicates if the company's board actively monitors or evaluates the obligations associated with health and safety (H&S) hazards, beyond just endorsing an H&S policy. Evaluate if there is unequivocal evidence of board or board committee supervision over the management of health and safety concerns
C7. Board background and skills	Details the professional expertise, talents, or age of each board member
C8. CEO (chief executive officer) board member	Expresses the board member of the CEO

TOPSIS model, is a hybrid methodology that combines the entropy technique with the TOPSIS method. The main purpose of this system is to determine the weight of each assessment indication by applying the entropy weight approach. It then employs a procedure to obtain the best solution for deciding the ranking of the assessment elements. The main goal of the entropy weight TOPSIS approach is to find the best solution in which all attribute values reach their best (or worst) levels within the alternative framework. An evaluation object is regarded to be the best if it is the closest to the best solution and the furthest from the worst solution, based on the relative distances between each evaluation object and these solutions. It is considered suboptimal if it does not have these properties. The concept of entropy weights the TOPSIS approach, making good use of information from the original dataset without placing any limits on the size of the sample. It has the advantages of little information loss and flexible functioning. Prior to presenting the approaches, we make the initial assumption that there are n alternatives $A=\{A_1,A_2,A_3,\ldots,A_n\}$, and m criteria $M = \{M_1,M_2,M_3,\ldots,M_m\}$, where, $i \in A$, $j \in M$, $i = \{1,2,3,\ldots,n\}$, $j=\{1,2,3,\ldots,m\}$.

The matrix $X' = (x'_{ij})$nxm in Eq. (3.1) is a decision matrix of $n \times m$. The weights of criteria M can be represented by a weight vector $W = \{w_1, w_2, w_3, \ldots, w_m\}$, which satisfies $\sum_{j=1}^{m} w_j = 1$.

$$X' = \begin{bmatrix} x'_{11} & x'_{12} & x'_{1m} \\ x'_{21} & \cdots & x'_{2m} \\ x'_{n1} & x'_{n2} & x'_{nm} \end{bmatrix} \tag{3.1}$$

The weight may be determined by the Shannon entropy weight technique, which is contingent upon the degree of dispersion of the data. Initially, we employ the Min-Max method to normalize the original choice matrix, which has dimensions of $n \times m$ $\{x'_{ij}\}$. Subsequently, we shift the standardized formula to the right by 0.001 units to simplify subsequent logarithmic computations.

$$x_{ij} = \frac{x'_{ij} - \min(x'_j)}{\max(x'_j) - \min(x'_j)} + 0.001, where\, i = 1,2,3,\ldots,n, and\, j = 1,2,3,\ldots,m. \tag{3.2}$$

Eq. (3.4) was employed to calculate the entropy value, denoted as e_j. Entropy quantifies the degree of dispersion of the data. As the data becomes more dispersed, the entropy number decreases, suggesting that the data contains a larger quantity of information. Indicating a reduction in the quantity of information contained in the data, the entropy value increases as the data becomes more concentrated.

$$r_{ij} = \frac{x_{ij}}{\sum_{i=1}^{n} x_{ij}}, i = 1,2,3,\ldots,n, and\, j = 1,2,3,\ldots,m. \tag{3.3}$$

$$e_j = -\frac{1}{\ln n}\sum_{i=1}^{n} r_{ij}, i = 1,2,3,\ldots,n \, and \, j = 1,2,3,\ldots,m. \tag{3.4}$$

Equation (3.5) is employed to determine the weight w_j.

$$w_j = \frac{1-e_j}{\sum_{j=1}^{m}\left(1-e_j\right)} \tag{3.5}$$

3.3 TOPSIS Model

Hwang and Yoon created the TOPSIS model to determine how close the alternatives are to the optimum answers. We determine how close the target options are to the ideal and anti-ideal solutions by calculating the Euclidean distance between them. The ideal solution is the one that has the best value for each assessment criterion, whereas the anti-ideal solution is the one that has the worst value for each evaluation criterion. The best option is chosen as the most appropriate alternative because it is the one that is most similar to the ideal solution and least similar to the anti-ideal solution. The first step is to equalize the positive and negative criteria in Eq. (3.6) in order to get rid of any differences in dimensions. It is important to point out that the Min-Max approach in Eq. (3.6) is necessary for the execution of standardization in several dimensions. This technique makes it possible to align gains and losses while

the entropy weight TOPSIS (Koemtzopoulos et al., 2025) is being developed.

$$positive: x_{ij}^{+} = \frac{x'_{ij} - \min\left(x'_j\right)}{\max\left(x'_j\right) - \min\left(x'_j\right)} \tag{3.6}$$

$$negative: x_{ij}^{-} = \frac{\max\left(x'_j\right) - x'_{ij}}{\max\left(x'_j\right) - \min in\left(x'_j\right)}$$

$$\min\left(x'_j\right) = \{\min_i x'_{ij} | 1 < i < n, 1 < j < m\}$$

$$\max\left(x'_j\right) = \{\max_i x'_{ij} | 1 < i < n, 1 < j < m\}$$

Then, the dimensionless standardized decision matrix x_{ij} is made by using normalized positive and negative criteria to make the initial choice matrix in Eq. (3.6), as shown in Eq. (3.7).

$$X' = \begin{bmatrix} x_{11} & x_{12} & \cdots & x_{1m} \\ x_{21} & \cdots & \cdots & x_{2m} \\ \vdots & \cdots & \cdots & \vdots \\ x_{n1} & x_{n2} & \cdots & x_{nm} \end{bmatrix}, where\, i = 1,2,3,\ldots,n\, and\; j = 1,2,3,\ldots,m. \quad (3.7)$$

Furthermore, the decision matrix in Eq. (3.8) is derived by multiplying each element $v_{ij} = w_j \times x_{ij}$, where $w_j = (w_1, w_2, w_3, \ldots, w_m$ is obtained from Eq. (3.5) and meets the condition $\sum_{j=1}^{m} w_j = 1$ and x_{ij} is generated using Eq. (3.7).

$$V = \begin{bmatrix} v_{11} & v_{12} & \cdots & v_{1m} \\ v_{21} & \cdots & \cdots & v_{2m} \\ \vdots & \cdots & \cdots & \vdots \\ v_{n1} & v_{n2} & \cdots & v_{nm} \end{bmatrix} = \begin{bmatrix} w_1 x_{11} & w_2 x_{12} & \cdots & w_m x_{1m} \\ w_1 x_{21} & w_2 x_{22} & \cdots & w_m x_{2m} \\ \vdots & \cdots & \cdots & \vdots \\ w_1 x_{n1} & w_2 x_{n2} & \cdots & w_m x_{nm} \end{bmatrix} \quad (3.8)$$

The positive ideal solution (PIS) is defined as the maximum value, and the negative ideal solution (NIS) is defined as the smallest value for each criterion in Eq. (3.9). To determine the distance of each alternative to the positive ideal solution (PIS) and the negative ideal solution (NIS), Eqs. (3.10) and (3.11) are employed.

$$PIS: P^{+} = \left\{ v_1^{+}, v_2^{+}, v_3^{+}, \ldots, v_m^{+} \right\} = \left\{ \left(\max_i v_{ij} | j \in M \right) \right\} \quad (3.9)$$

$$NIS: P^{-} = \left\{ v_1^{-}, v_2^{-}, v_3^{-}, \ldots, v_m^{-} \right\} = \left\{ \left(\min_i v_{ij} | j \in M \right) \right\}$$

$$d_i^{+} = \sqrt{\sum_{j=1}^{m} (v_{ij} - v_j^{+})^2}, i = 1,2,3,\ldots,n, and\; j = 1,2,3,\ldots,m. \quad (3.10)$$

$$d_i^{-} = \sqrt{\sum_{j=1}^{m} (v_{ij} - v_j^{-})^2}, i = 1,2,3,\ldots,n, and\; j = 1,2,3,\ldots,m. \quad (3.11)$$

Ultimately, calculate the coefficient of relative closeness (RC).

$$RC_i = \frac{d_i^{-}}{d_i^{-} + d_i^{+}}, i = 1,2,3,\ldots,n \quad (3.12)$$

3.4 Results

The research established a decision matrix to categorize the analysis into three fundamental components. The following subsections delineate the various stages and their corresponding procedures: The preliminary stage utilizes the N1 method to

Table 3.2 Weight vector of the criteria for FY2024

Criterion	C1	C2	C3	C4	C5	C6	C7	C8
ej	0.9870	0.9775	0.9904	0.9899	0.9899	0.9977	10.000	0.9901
d = 1-ej	0.0130	0.0225	0.0096	0.0101	0.0101	0.0023	0.0000	0.0099
Wj	0.1674	0.2902	0.1241	0.1299	0.1304	0.0300	0.0000	0.1279

standardize the data inside the decision matrix. This technique is designed to standardize the data for enhanced comparability. The subsequent phase of the research included assigning weight to each criterion and constructing the weighted normalized matrix, as detailed in Sect. 3.2 of this study. This research was done using the entropy weight TOPSIS approach. Table 3.2 delineates the weights allocated to each evaluated criterion for FY2024.

The governance pillar significantly influences the greenwashing ratings of European financial institutions when the entropy weight and TOPSIS methodologies are used together. In fact, the last 2 years seem to constitute a governance bubble characterized by rapidly escalating expectations for 'greenness,' while the governance structures inside firms are failing to adapt. In this context, it is crucial to acknowledge the connection between greenwashing and governance and to establish robust processes to ensure that corporations can substantiate their environmental, social, and governance (ESG) assertions. Emphasizing the "G" in ESG is a crucial method for risk mitigation and enhancing stakeholder confidence. Let us examine how a corporation may inadvertently partake in misleading tactics while asserting its commitment to environmental sustainability. This action is not always malevolent or purposeful. In several instances, the auditors lacked the requisite expertise to substantiate their assertions. Numerous corporations claim adherence to self-imposed norms, misleading customers into believing they engage in ecologically sustainable operations due to their status as well-known brands. Nevertheless, the dependability of ESG data presented in companies' sustainability reports is often unaudited. Unreliable ESG information may impede the incorporation of ESG aspects into investment choices due to a company's greenwashing practices. Recognizing the proliferation of greenwashing increases the potential for legal complications for corporations.

The misunderstanding of ESG standards and the absence of a cohesive framework often lead organizations to make insincere pledges instead of actually embracing sustainable practices. It is essential to provide a framework for governance experts to develop "greenwash-proof" enterprises. This framework must include comprehensive due diligence procedures, autonomous audits, and explicit directives for ESG reporting and disclosure. Organizations must implement internal controls to ensure the precision and dependability of their ESG assertions. Furthermore, regulatory agencies and industry organizations are essential in formulating industry standards and overseeing compliance. Implementing this methodology would enable governance experts to verify that their firms are truly dedicated to sustainability, thus circumventing the risks of greenwashing. This strategy would cultivate trust among stakeholders and facilitate informed investment choices (Koemtzopoulos et al., 2025; Ragazou et al., 2024; Zournatzidou, Sklavos, Ragazou, & Sariannidis, 2024).

Furthermore, the results demonstrate that the greenwashing score is significantly affected by board size, which is the second most critical element. The relationship between board size and greenwashing may seem debatable, since it is grounded in the existing literature on corporate governance and environmental performance. Some experts contend that bigger boards result in increased inefficiencies in the interactions between a company's insiders and outsiders. Conversely, others contend that enhanced oversight reduces the impact of individual members on the whole board. A larger board may have committees explicitly tasked with addressing sustainability concerns. Moreover, a significant association exists between the size of the board and its environmental effectiveness. This indicates an inverse relationship between greenwashing and board size.

Furthermore, the board's efficacy should serve as a stimulus for improving the quantity, transparency, and quality of environmental disclosures. Nevertheless, the present analysis indicates that this feature ranks as the third most significant factor in the prevalence of greenwashing. Independent board members are often seen as more effective in supervising CEOs, less susceptible to managerial influence, and more inclined to prioritize the interests of shareholders. Nevertheless, having insiders supervise the board might be advantageous, since insider control more effectively leverages proprietary information. Considering the potential motives for seemingly independent directors to collaborate with insiders, investors have often surpassed the conventional expectations for board efficacy (Tzitiridou-Chatzopoulou et al., 2024; Zournatzidou, Mallidis et al. 2024). Regulatory constraints impede independent directors from promoting environmental sustainability, but normative pressures foster such advocacy. This may explain the inverse correlation between the board's performance and the amount of information the firm willingly discloses. This may explain why regulatory and normative constraints have disparate influences on independent directors' efforts to enhance the sustainability performance of their enterprises. This indicates a correlation between board independence and greenwashing, since insufficient scrutiny from external directors may increase the likelihood of corporations using greenwashing tactics (Brunsson, 1993; Koch & Denner, 2025).

3.5 Discussion

The increasing emphasis on sustainability is a positive trend, as companies are rapidly incorporating environmental considerations into their strategies and operations. Consumers are selecting environmentally responsible products, governments are enforcing more stringent environmental regulations, and investors are requesting ESG disclosures. In an effort to remain competitive, some organizations have also fabricated or exaggerated their sustainability accomplishments as a result of this heightened emphasis. This phenomenon, which is referred to as "greenwashing," is not only detrimental to the company and its constituents, but it is also deceptive. Various types of greenwashing exist, including minor exaggerations and blatant fabrications. Companies may portray their sustainability initiatives in a positive perspective while neglecting to disclose the adverse environmental consequences of other aspects of their operations. For instance, a company may emphasize its utilization of recycled materials in packaging while disregarding the fact that its supply chain emits substantial carbon emissions. This selective reporting creates a deceptive impression of sustainability and misleads stakeholders into believing that the company is more environmentally responsible than it honestly is.

New governance challenges are presented to boards of directors by the proliferation of greenwashing. Directors are responsible for guaranteeing that their organizations not only adhere to environmental regulations but also make sustainability claims that are both transparent and truthful (Li et al., 2024; Wang et al., 2024, 2024; Wang & Shen, 2024). As regulators and consumers are increasingly holding companies accountable for misleading claims, failure to comply with this requirement can lead to financial risks and reputational damage. As custodians of corporate integrity, boards are obligated to supervise sustainability reporting and guarantee that the claims made by their organizations are substantiated by verifiable data and consistent with genuine environmental advancements.

The potential of specific corporate governance mechanisms to either facilitate or impede greenwashing is the subject of this paper. We have implemented an integration of the entropy weight and TOPSIS multicriteria decision-making methods to address the research objective of the current study. We initially propose a novel set of metrics to accurately represent the enigmatic phenomenon of greenwashing. Consequently, we can illustrate the nature of the relationships between specific board characteristics and greenwashing behaviors. It is unclear how the intensity of greenwashing is correlated with the number of directors in the context of board size. Nevertheless, the notion that a larger board size implicitly guarantees a resolution to greenwashing can be disproven. Greenwashing behaviors seem to be positively correlated with the efficacy of boards. This may seem to be a contradiction; however, it is in accordance with the idea that independent directors can increase their reputation with stakeholders and secure additional employment opportunities by improving the environmental credentials of the company they represent.

Additionally, the results revealed that the greenwashing score of financial institutions in Europe can be influenced by the magnitude of the board. According to certain scholars in the literature, the increased agency costs that are linked to larger boards are a consequence of the increased inefficiencies in the interaction between insiders and outsiders of firms. Similar to this, the literature poses inquiries regarding the causal relationship between increased performance and board size. We may be more susceptible to greenwashing if we neglect to incorporate enhanced monitoring into larger committees. Conversely, some scholars argue that larger boards result in improved monitoring, which in turn diminishes the influence of individual (inside) members on the board, thereby reducing agency costs. In order to resolve sustainability concerns, organizations with larger boards frequently establish specific committees, which leads to generally superior environmental performance. The board size can serve as a viable solution to greenwashing, as our research validates.

References

Brunsson, N. (1993). Ideas and actions: Justification and hypocrisy as alternatives to control. *Accounting, Organizations and Society, 18*, 489–506. https://doi.org/10.1016/0361-3682(93)90001-M

Gomes, M., Marsat, S., Peillex, J., & Pijourlet, G. (2024). Does religiosity influence corporate greenwashing behavior? *Journal of Cleaner Production, 434*. https://doi.org/10.1016/J.JCLEPRO.2023.140151

Koch, T., & Denner, N. (2025). Different shades of green deception. Greenwashing's adverse effects on corporate image and credibility. *Public Relations Review, 51*. https://doi.org/10.1016/j.pubrev.2024.102521

Koemtzopoulos, D., Zournatzidou, G., & Sariannidis, N. (2025). Can cryptocurrencies be green? The role of stable coins toward a carbon footprint and sustainable ecosystem. *Sustainability, 17*, 483. https://doi.org/10.3390/SU17020483

Lagoarde-Ségot, T. (2024). Greenwashing and sustainable finance: An approach anchored in the philosophy of science. *Current Opinion in Environment Sustainability, 66*. https://doi.org/10.1016/J.COSUST.2023.101397

Li, J. (2024). Controlling shareholders' stock pledges and greenwashing–evidence from China. *Finance Research Letters, 69*. https://doi.org/10.1016/j.frl.2024.106227

Li, T., Shu, X., & Liao, G. (2024). Does corporate greenwashing affect investors' decisions? *Finance Research Letters, 67*. https://doi.org/10.1016/j.frl.2024.105877

Li, Z., Miao, S., & Xu, L. (2024). Digital transformation and environmental, social, and governance greenwashing: Evidence from China. *Journal of Environmental Management, 365*. https://doi.org/10.1016/J.JENVMAN.2024.121460

Liu, Z., & Li, X. (2024). The impact of bank fintech on ESG greenwashing. *Finance Research Letters, 62*. https://doi.org/10.1016/j.frl.2024.105199

Peng, Q., & Xie, Y. (2024). ESG greenwashing and corporate debt financing costs. *Finance Research Letters, 106012*. https://doi.org/10.1016/j.frl.2024.106012

Ragazou, K., Zournatzidou, G., Sklavos, G., & Sariannidis, N. (2024). Integration of circular economy and urban metabolism for a resilient waste-based sustainable urban environment. *Urban Science, 8*, 175. https://doi.org/10.3390/URBANSCI8040175

Tzitiridou-Chatzopoulou, M., Orovou, E., & Zournatzidou, G. (2024). Digital training for nurses and midwives to improve treatment for women with postpartum depression and protect neonates: A dynamic bibliometric review analysis. *Healthcare, 12*, 1015. https://doi.org/10.3390/HEALTHCARE12101015

Wang, H., & Shen, H. (2024). Self-restraint or external supervision: Green bond issuance and greenwashing. *Research in International Business and Finance, 70*. https://doi.org/10.1016/J.RIBAF.2024.102402

Wang, Q., Ma, Z., Zhao, J., & Shu, G. (2024). State shareholding in privately-owned firms and greenwashing: Evidence from China. *Finance Research Letters, 62*. https://doi.org/10.1016/J.FRL.2024.105176

Wang, Y., Xing, C., & Zhang, L. (2024). Is greenwashing beneficial for corporate access to financing? Evidence from China. *Finance Research Letters, 65*. https://doi.org/10.1016/J.FRL.2024.105538

Xue, K., Chen, X., Sun, P., & Li, Y. (2024). Supply chain network centrality and corporate greenwashing behavior. *Finance Research Letters, 70*. https://doi.org/10.1016/j.frl.2024.106357

Yin, L., & Yang, Y. (2024). How does digital finance influence corporate greenwashing behavior? *International Review of Economics and Finance, 93*, 359–373. https://doi.org/10.1016/j.iref.2024.05.014

Zhao, X., Huang, X., Liu, F., & Pan, L. (2024). Executive power discrepancy and corporate ESG greenwashing. *International Review of Financial Analysis, 96*. https://doi.org/10.1016/j.irfa.2024.103533

Zournatzidou, G. (2024). Evaluating executives and non-executives' impact toward ESG performance in banking sector: An entropy weight and TOPSIS method. *Administrative Sciences, 14*, 255. https://doi.org/10.3390/ADMSCI14100255

Zournatzidou, G., Sklavos, G., Ragazou, K., & Sariannidis, N. (2024). Anti-competition and anti-corruption controversies in the European financial sector: Examining the anti-ESG factors with entropy weight and TOPSIS methods. *Journal of Risk and Financial Management, 17*, 492. https://doi.org/10.3390/JRFM17110492

Zournatzidou, G., Mallidis, I., Farazakis, D., & Floros, C. (2024). Enhancing bitcoin price volatility estimator predictions: A four-step methodological approach utilizing elastic net regression. *Mathematics, 12*, 1392. https://doi.org/10.3390/MATH12091392

Examining the Relationship Between ESG Standards and Corporate Corruption: Greenwashing's Substantial Impact

4

4.1 Greenwashing: A Type of Corruption That Impacts Climate Risk

The literature contains substantial research on greenwashing in the field of sustainable finance. Wang et al. (2024) offer a thorough examination of the factors that contribute to greenwashing, emphasizing the significance of both internal and external variables (Wang et al. 2024). They contend that businesses indulge in greenwashing for a variety of reasons, such as to enhance their company image, comply with legislative obligations, and respond to pressure from stakeholders. Moreover, Huang et al. (2024) investigate the rationale behind and impediments to the implementation of environmentally friendly supply chain management strategies, a critical domain in which greenwashing may manifest (Huang et al., 2024). They discover that despite the substantial incentives for businesses to implement environmentally favorable practices, there are also significant challenges, such as expenses and a lack of legitimacy.

Furthermore, a significant area of research has been the examination of the impact of greenwashing on a company's financial success and reputation. For instance, it has been determined that greenwashing may diminish a company's reputation, which may result in a decline in consumer confidence and the possibility of financial penalties. Conversely, greenwashing may occasionally serve to improve a company's financial performance by appealing to investors and consumers who prioritize environmental sustainability. Nevertheless, it is anticipated that this effect will be transient as consumers become more cognizant of the organization's dishonest practices.

Another significant motif in the literature is the involvement of stakeholders in greenwashing. Stakeholders, particularly consumers and investors, play a critical role in the promotion or prevention of greenwashing. There are perceptions and responses of stakeholders to greenwashing that can have a substantial impact on a

K. Ragazou et al., *Transparency and Corporate Washing in the Age of Sustainability*, Management for Professionals,
https://doi.org/10.1007/978-3-031-96821-1_4

company's strategy. Also, consumers' perceptions of a company's corporate social responsibility initiatives may significantly influence their reactions to greenwashing. They were in favor of this concept. Additionally, this study can offer valuable insights into the measures that can be implemented to combat greenwashing. For example, additional regulations and standards in sustainable financing are required to prevent greenwashing. Further research in this field emphasizes the significance of education and awareness in the fight against greenwashing (Teti et al., 2024; Zhou et al., 2024; Oguntona, 2024; Torelli et al., 2020).

Another significant area of investigation is the connection between sustainable development and greenwashing. Peng and Xie (2024) posit that "greenwashing" may impede sustainable development by diverting resources from sustainable endeavors. He contends that greenwashing may result in a misunderstanding of sustainability, which can subsequently result in the inefficient use of resources. This concept demonstrates that greenwashing can result in skepticism and mistrust among consumers, thereby restricting the implementation of sustainable practices. There is a variety of sector-specific studies that provide evidence of greenwashing in the literature. Thus, previous studies have underscored the necessity of implementing measures that are specific to each industry to combat the widespread prevalence of greenwashing.

The relationship between sustainable finance and greenwashing is a multifaceted issue that is influenced by a variety of theoretical foundations, methodologies, and perspectives. It is imperative to comprehend the correlation between sustainable finance and technological innovations such as fintech. Digital financial solutions have the potential to reduce greenwashing activities by increasing transparency and safeguarding consumers, as per Vergara and Agudo (2021) (Chueca Vergara et al., 2021). Moodaley and Telukdarie (2023) emphasize the capacity of artificial intelligence to evaluate sustainability reports as indicators of greenwashing, thereby providing a novel approach to enhance the credibility of corporate environmental disclosures (Moodaley & Telukdarie, 2023). Artificial intelligence is employed to further investigate this synergy.

4.2 Materials and Methods

Section 4.3 uses bibliometrics to clarify the findings. Bibliometric analysis is essential for examining the citation metrics and other statistical characteristics of research articles. Published research statistics can be analyzed using bibliometric methods. Researchers assert that bibliometric analysis is a legitimate approach for analyzing large scientific datasets. Advancements in disciplines influence academic domains. Bibliographic analysis has enabled researchers to assess published works, identify emerging research trends, and understand collaborative dynamics. Furthermore, we must interrogate the database using suitable search terms. Table 4.1 illustrates the compilation of these articles using keyword research methodologies. Initially, we used keyword research to conduct an extensive search of Scopus's scientific article titles and terms from 2012 to 2024. These databases enhance the effectiveness of

Table 4.1 Keyword search formula. Source: Own elaboration

Step	Keyword search
1	((“corruption” AND “ESG”))
2	((“corruption” OR “fraud”) AND (“ESG”))
3	((“corruption” OR “fraud” OR “controversies”) AND (“ESG”))
4	((“corruption” OR “fraud” OR “controversies”) AND (“corporate washing”) AND (“ESG”))
5	((“corruption” OR “fraud” OR “controversies”) AND (“corporate washing” OR “greenwashing”) AND “ESG”))
6	((“corruption” OR “fraud” OR “controversies”) AND (“corporate washing” OR “greenwashing” OR “bluewashing”) AND (“ESG”))
7	((“corruption” OR “fraud” OR “controversies” OR “money laundering”) AND (“corporate washing” OR “greenwashing” OR “bluewashing”) AND (“ESG”))
8	((“corruption” OR “fraud” OR “controversies” OR “money laundering”) AND (“corporate washing” OR “greenwashing” OR “bluewashing” OR “pinkwashing”) AND (“ESG”))
9	((“corruption” OR “fraud” OR “controversies” OR “money laundering”) AND (“corporate washing” OR “greenwashing” OR “bluewashing” OR “pinkwashing”) AND (“ESG” OR “sustainability” OR “sustainable development” OR “circular economy”) AND (“climate risk”))
10	((“corruption” OR “fraud” OR “controversies” OR “money laundering”) AND (“corporate washing” OR “greenwashing” OR “bluewashing” OR “pinkwashing”) AND (“ESG” OR “sustainability” OR “sustainable development” OR “circular economy”) AND (“climate risk”)) AND (LIMIT-TO (PUBSTAGE, “final”) OR LIMIT-TO (PUBSTAGE, “aip”)) AND (LIMIT-TO (SRCTYPE, “j”))

bibliographic analyses and literary studies; thus, we selected them (Khan et al., 2023; Elshabshiri et al., 2025).

Authors, publications, and institutions are used to assess the research efficacy. We illustrate the bibliographic connection using the performance data and VOSviewer. We then provide data for analysis, emphasizing scientific mapping and performance assessment, which are two essential bibliometric elements. This research assesses bibliometrics using all three methodologies. Thematic maps and academic research topics are created using Biblioshiny's network analysis and performance-evaluation metrics. Scientific mapping, network analysis, and performance assessment instruments enable researchers to assess their studies and enhance their visualization. We evaluate the foremost authors, institutions, and journals using the h-index, total citations, and publishing metrics. We identify phrases with analogous subjects via a clustering methodology (Khan et al., 2023).

The R software offers sophisticated data analysis functionalities for this research. Bibliometrix is an open-source software that analyzes scientific literature for the purpose of science mapping. This instrument facilitates researchers' quantitative bibliometric assessments. Bibliographic data from Scopus are accurately imported using the R-based RStudio version 3.6.0 or above. Diverse scientific fields have meticulously examined the significance of R and its libraries. The official website states that this study conducted a bibliometric analysis using the Biblioshiny online tool, which is a component of the Bibliometrix program owing to its intuitive interface. This research's software tools are accessible to everyone, particularly to scholars. These gadgets can detect trends and deficiencies, as shown in previous research and databases. Individuals without computer proficiency or clustering knowledge may use Mendeley and VOSviewer to categorize articles. Data visualization and analysis require the evaluation of three hierarchical metrics: sources, authors, and documents. There are three categories of organized knowledge frameworks: an intellectual framework using network analysis and historical methodologies, a social framework employing connection network techniques, and a conceptual framework including network, factorial, and thematic mapping.

4.3 Data

The data utilized in this investigation were sourced from the Scopus database. Scopus is the most exhaustive database of peer-reviewed scientific literature, as provided by Elsevier BV, which is headquartered in the United States. There are more than 23,000 works in this compilation that have been the subject of extensive scientific research by international publishers. It is the goal of this methodology to comprehend the theoretical underpinnings of the procedures in this field. This methodology is designed to identify research deficiencies and generate new opportunities for additional research. In addition, we implemented the PRISMA application. The PRISMA method (Fig. 4.1) was also employed to collect information, which enabled us to determine whether it was feasible to conduct bibliometric analyses and identify pertinent scientific articles. The significance and trends of ESG in relation to greenwashing have been the focus of a series of studies that exclusively address this subject. A comprehensive screening process was conducted for this investigation. To conduct an additional evaluation of the research, we restrict the document categories exclusively to research papers. To guarantee the consistency and uniformity of the articles investigated in this investigation, we implement this procedure (Page et al., 2021; Smith et al., 2024; Galletta et al., 2024).

The PRISMA flow diagram in Fig. 4.1 illustrates the primary methodologies employed to systematically select publications for the bibliometric research. The search query identified a total of 846 sources in the collection. Consequently, we restricted the number of articles, resulting in a reduction of the total to 653. We meticulously evaluated 431 research publications, excluding those that were too general or lacked a distinct relationship, as they were unsuitable for the current study. The primary objective of this investigation is to evaluate the impact of greenwashing on ESG reporting. We discovered that the titles or keywords of the specific sources we selected do not adequately capture the characteristics and specifics of the topic under investigation after conducting a thorough examination of the papers. The search parameters were modified to focus on research that was pertinent to the current investigation and to eliminate irrelevant references. Using this filtration procedure, a total of 319 academic publications were identified and incorporated into the bibliometric analysis (Page et al., 2021; Smith et al., 2024; Galletta et al., 2024).

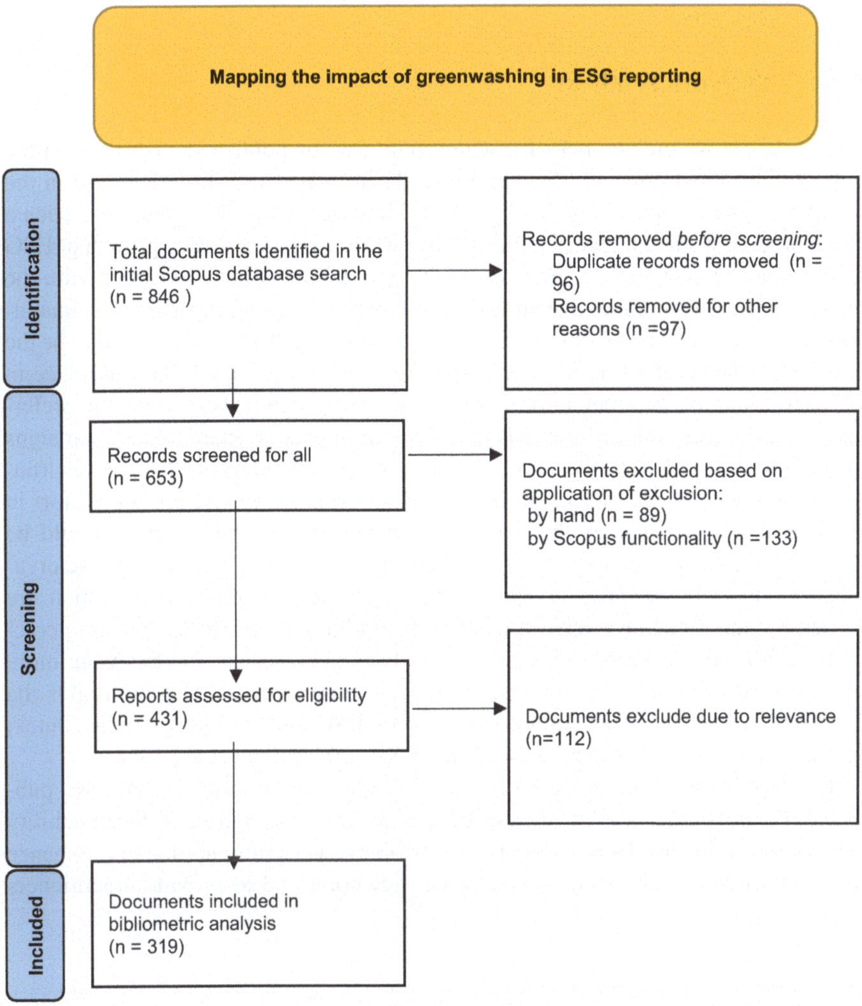

Fig. 4.1 PRISMA flow diagram

In addition, the PRISMA methodology established the following inclusion and exclusion criteria to simplify the source selection process: (i) publication date; (ii) correlation between ESG and greenwashing; (iii) language of publication (only English-language research studies are considered); and (iv) geographic factors, which encompass specific regions, states, countries, or populations. Therefore, each of these articles was incorporated into the bibliometric analysis.

4.4 Results

4.4.1 Network Analysis

Figure 4.2 shows the number of research publications published each year on the impact of greenwashing in ESG reporting. There was a very little increase in the number of publications from 2012 to 2021. However, since 2022, there has been a large increase in the number of publications on this topic. The steady growth of ESG investment saw a significant increase in 2013 and 2014, which coincided with the release of the first research that showed a link between strong company sustainability performance and positive financial results. However, 2018 is considered to be the most important year for ESG. The corporate proxy season for ESG took place in 2018, resulting in the emergence of several evolving trends that clarify the preferences of investors and are anticipated to have an impact on shareholder campaigns in the next year. Shareholders are becoming more and more involved in environmental, social, and governance problems. Given the growing interest of investors in ESG problems, companies that have not yet engaged in these topics should be encouraged to start reaching out to and interacting with key investors. These organizations include stewardship, governance, and proxy voting teams, which are owned by numerous investors, in addition to traditional investor relations interactions with fund managers and securities analysts. As a result, scholars were more interested in ESG, and this interest grew even more in 2024. This highlights the necessity to investigate the connection between ESG and corruption in the context of greenwashing (Li & Chen, 2024; Peng & Xie, 2024; Zhao et al., 2024).

Figure 4.3 shows the breadth of greenwashing and the journals that have published the most articles in the area of ESG reporting. The Journal of Sustainability is at the top of this list. Its objective is to distribute research articles that may enhance the knowledge of ESG from a worldwide viewpoint and to provide practitioners

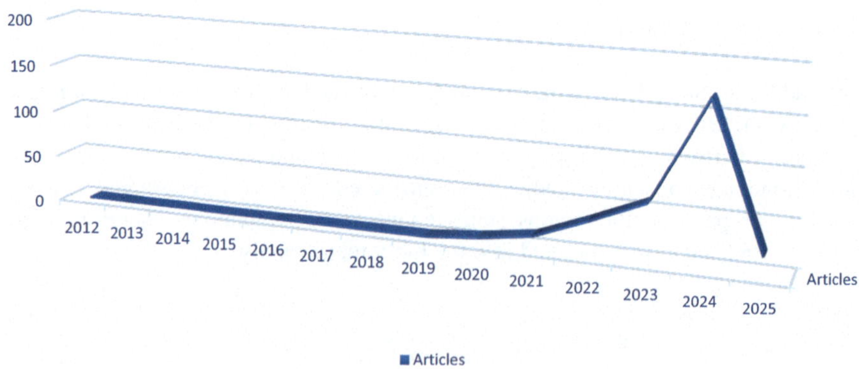

Fig. 4.2 Annual growth of publications in the field. Source: Scopus/R Studio

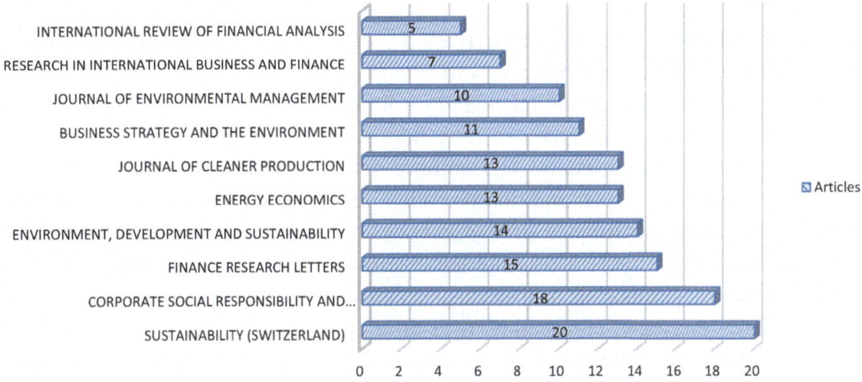

Fig. 4.3 Most impactful sources in the field. Source: Scopus/R Studio

with management insights and suggestions that can assist them in developing and improving ESG strategies. The journal *Corporate Social Responsibility and Environmental Management* (Wang et al. 2024, 2024, 2024) is in second place on the list. Its purpose is to add to the current body of knowledge by publishing articles that investigate the connection between ESG elements and their function in reducing the risks that are linked to greenwashing techniques in the financial industry. Consequently, the topic of discussion is the insufficient amount of research that has been conducted on the relationship between the dangers of greenwashing and the implementation of ESG practices by financial institutions in order to reduce such dangers. Furthermore, the main objective of this project is to investigate the environmental, social, and governance (ESG) measures that might help mitigate the adverse effects of greenwashing. The study will also give regulators recommendations on how to prevent financial institutions from participating in dishonest environmental practices (Li & Chen, 2024; Peng & Xie, 2024; Zhao et al., 2024).

The Journal of Finance Research Letters is the third most prolific source of papers on this topic. This journal has published one of the most recent and significant research projects, which highlights a different relationship between ESG reporting and greenwashing. In particular, it examines ESG greenwashing in fintech businesses. This study, which is called "The Impact of Bank Fintech on ESG Greenwashing," looks at how bank fintech affects ESG greenwashing. In doing so, it adds to the current body of information on ESG greenwashing and the literature on the financial implications of fintech in banks. In addition, this study presented a technique for assessing firm-level bank fintech by using text analysis to identify fintech patents among all banks' patents and loan-by-loan data to assign a firm-level weight to bank fintech. Additionally, this study makes another important addition to the topic being investigated by using the authors in a moderating effect model to analyze the basic processes that affect the role of bank fintech in ESG greenwashing. Furthermore, the top five sources with the most relevant papers on the topic include the Environment, Development, and Sustainability journal (14 publications) and the Energy Economics journal (13 publications).

The most important and impactful research publications on this topic are summarized in Table 4.2. The first research article on the list is "Greenwashing in environmental, social, and governance disclosures," which was written by Yu, Van Luu, and Chen in 2020. This article takes a close look at the ways in which companies may reduce the amount of greenwashing that occurs in the environmental, social, and governance (ESG) aspects. The authors define "greenwashers" as firms that seem to be open and honest and that provide a lot of information on their environmental, social, and governance (ESG) practices but that do not really perform well in these areas. The authors determine the extent to which large-cap companies use greenwashing by computing peer-relative greenwashing scores using a dataset that includes 1925 large-cap companies from different countries. The most important findings of this research were that greenwashing behavior in ESG dimensions can be discouraged by (i) independent directors, (ii) institutional investors, (iii) influential public interests through a less corrupt country system, and (iv) being cross-listed.

Furthermore, the results indicate that the two governance variables at the business level are the most efficient in reducing the deceptive disclosures of enterprises on ESG dimensions. The research article "Stakeholder legitimacy in firm greening

Table 4.2 Most impactful research publications in the field. Source: Scopus/R Studio

Paper	Total citations	TC per year	Normalized TC
Greenwashing in environmental, social, and governance disclosures (Yu et al., 2020)	405	67.50	3.38
Corporate social responsibility governance, outcomes, and financial performance (Wang & Sarkis, 2017)	388	43.11	1.00
CSR Performance and the Readability of CSR Reports: Too Good to be True? (Wang et al., 2018)	230	28.75	1.00
ESG did not immunize stocks during the COVID-19 crisis, but investments in intangible assets did (Demers et al., 2021)	211	42.20	4.81
Is corporate social responsibility reporting a tool of signaling or greenwashing? Evidence from the worldwide logistics sector (Uyar et al., 2020)	159	26.50	1.33
Green financial system regulation shock and greenwashing behaviors: Evidence from Chinese firms (Zhang, 2022b)	155	38.75	4.63
Stakeholder legitimacy in firm greening and financial performance: What about greenwashing temptations? (Lee & Raschke, 2023)	134	44.67	6.31
Are firms motivated to greenwash by financial constraints? Evidence from global firms' data (Zhang, 2022a)	119	29.75	3.55
What Do We Mean by Sustainable Finance? Assessing Existing Frameworks and Policy Risks (Migliorelli, 2021)	102	20.40	2.33
Corporate uptake of the Sustainable Development Goals: Mere greenwashing or an advent of institutional change? (Lashitew, 2021)	100	20.00	2.28

and financial performance: What about greenwashing temptations?" was published by Lee and Raschke in 2023. It has the greatest normalized Total Citations (TC), even though it has the most citations overall. Specifically, the normalized TC index for this research is 6.31, which indicates that it has a dynamic relationship with the other studies on the list.

4.4.2 Bibliometric Analysis

Figure 4.4 shows the Sankey diagram, also known as the three-field plot, of the references that were mentioned. This diagram includes the authors, the nations of the authors, and phrases. Due to the legal, financial, and reputational dangers associated with it, a lot of businesses and investors are worried about greenwashing. As a result, it seems sensible that ESG raters would consider it a risk exposure issue and a point of contention in their ratings. The Sankey diagram in this study shows that greenwashing is a major issue for many firms, and investors are concerned about it due to the legal, financial, and reputational concerns involved. As a result, it seems sensible that ESG raters would consider it a risk exposure issue and a point of contention in their ratings. The graph may also provide a more extensive explanation of the importance of ESG ratings and the need for firms and ESG raters to understand the dangers of greenwashing and include relevant information in their ratings (Jiang et al., 2024; Peng & Kong, 2024).

Because stakeholders of businesses rely on ESG ratings to provide an accurate picture of the health, dangers, and prospects of enterprises while avoiding greenwashing concerns, this might have an impact on the three key components of their expectations of ESG raters. More specifically, ESG may help organizations assess their ESG policies and performance in order to avoid misleading stakeholders and investors by carefully examining and analyzing claims for veracity. In addition,

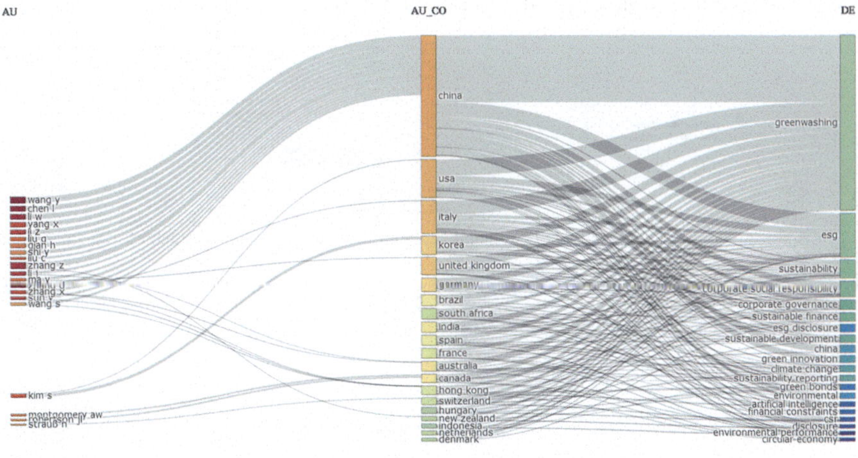

Fig. 4.4 Three-field plot. Source: Scopus/R Studio

ESG promotes businesses to improve their performance in order to increase investor confidence by fostering corporate responsibility and transparency. However, the legitimacy of the ranking is put at risk in the presence of neglected greenwashing. In addition, businesses may provide investors with trustworthy information by adopting ESG. This helps investors make educated choices by allowing them to match their portfolios with firms that are sustainable and responsible (Zhao et al., 2024).

Furthermore, the data highlight the importance of green innovation, as opposed to greenwashing, which may help companies maintain their ESG reporting process. Over the last five years, it has been shown that ESG strategies may motivate corporations to explore green innovation and push them to include environmental protection measures in their growth plans. This gives them the ability to improve their environmental protection via market signals. On the other side, greenwashing is when companies make false or misleading statements about their environmentally friendly actions in order to create a green image (Li & Xiao, 2025). This disconnect between environmental claims and environmental activities might have a negative impact on their green innovation strategies. Businesses that participate in green innovation may make a real contribution to environmental preservation, in addition to lowering the incentives for greenwashing. This creates a "win-win" situation for both corporate growth and sustainability. In addition, green innovation may help minimize greenwashing by allowing firms to take risks, which in turn creates a more stable financial climate. In addition, it may stabilize the governance framework by increasing the variety of administrators in terms of their gender and environmental setting. Heterogeneity tests have shown that this method is most successful in firms that have high levels of pollution and improved economic performance.

Figure 4.5 presents a thematic map that facilitates the identification of the intellectual foundations of the subject matter. A thematic map is designed to facilitate a deeper understanding of the current conditions of the field and its potential for future sustainability. This study is advantageous as it offers researchers and stakeholders valuable insights into the potential for future advancements in research

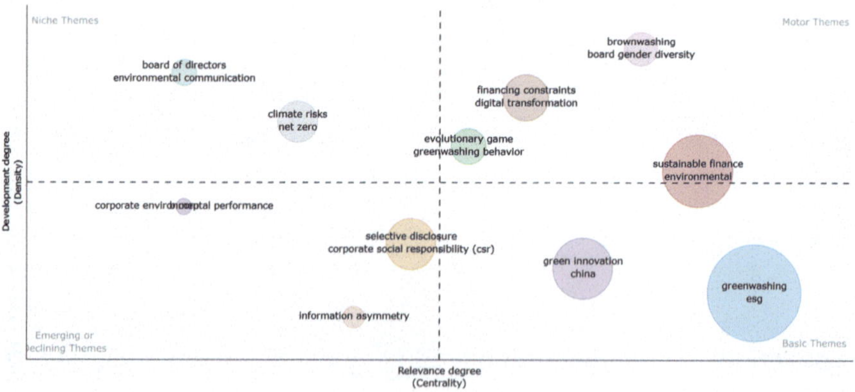

Fig. 4.5 Thematic analysis. Source: Scopus/Biblioshiny

themes related to a specific topic. Thematic analysis is a methodological approach that entails investigating the relationships among clusters of authors' keywords in order to identify the underlying themes. These elements are distinguished by their centrality and density. The vertical axis represents density, whereas the horizontal axis indicates centrality. Density quantifies the degree of cohesion among the nodes, whereas centrality evaluates the breadth of connections across various themes. The significance and profundity of specific subjects are determined by these two characteristics. As the number of connections among the nodes within the thematic network increases, a node achieves greater centrality and significance. This is the most pivotal function of a network. Similarly, the cohesiveness of a node, which serves as an indicator of the density within a research domain, significantly impacts its capacity for persistence and growth.

Furthermore, Fig. 4.5 illustrates the thematic map that delineates the impact of greenwashing on Environmental, Social, and Governance (ESG) reporting. The map is predominantly segmented into four quadrants, designated as Q1–Q4. The principal elements are illustrated in the upper-right quadrant (Q1), while the essential topics are delineated in the lower-right quadrant (Q4). Highly specialized themes are located in the upper left quadrant (Q2), whereas emergent or declining themes are positioned in the lower left quadrant (Q3). The findings from the thematic map indicate a substantial correlation between corporate governance and greenwashing, which has implications for Environmental, Social, and Governance (ESG) reporting. The critical function of the board of directors in the implementation of greenwashing is illustrated by the specific concerns identified in the thematic map.

An analysis of the correlation between corporate governance and greenwashing may assist in the identification of Corporate Social Responsibility (CSR) committees, thereby improving their environmental performance and reducing the prevalence of greenwashing. This analysis of the CSR committee's indirect involvement may enhance the comprehension of board attributes and the phenomenon of greenwashing. A board possessing a diverse array of skills is considered to strengthen internal controls, thereby mitigating the challenges associated with corporate social responsibility (CSR) and facilitating a better understanding of the dissemination of environmental information. The thematic map elucidates the relationship between environmental communication and greenwashing concerning the specific issues at hand. The term "greenwash" refers to a spectrum of deceptive communications aimed at fostering an unduly positive perception among stakeholders regarding an organization's environmental initiatives (Hussain et al., 2023; Zhao et al., 2022). The proliferation of claims regarding corporate social responsibility impedes stakeholders' ability to differentiate between organizations that genuinely engage in ethical practices and those that merely offer superficial endorsements of a sustainable development framework. It is imperative to comprehend and evaluate the influence of diverse misleading environmental communications on the public's perceptions of corporate environmental accountability and greenwashing. This is illustrated in a thematic map that analyzes the issue from the perspectives of legitimacy and signaling theory.

Furthermore, the findings of the co-occurrence analysis, which employed the most relevant and significant terms pertaining to the subject matter under examination, are illustrated in Fig. 4.6. The purple cluster appears to be the most prominent among the clusters generated in this study. This cluster serves as a quintessential example of the essential function of Environmental, Social, and Governance (ESG) criteria in mitigating the risk of greenwashing. The importance of transparency in the fight against greenwashing and the various dimensions of Environmental, Social, and Governance (ESG) considerations is underscored by the widespread usage of the related terminology. Transparency functions as a conduit between organizations and their constituents, providing a comprehensive comprehension of the company's ESG activities. Moreover, ESG transparency entails the organization's dedication to sustainable and ethical business practices, along with the disclosure of its activities. Transparency is essential for ESG to guarantee that corporations are held accountable for their actions. Transparency refers to the practice of revealing relevant information regarding a company's governance procedures, environmental effects, and social accountability. This information is indispensable for stakeholders, including investors, consumers, and laborers, to make well-informed decisions. The adoption of sustainable practices by corporations is influenced by the transparency of

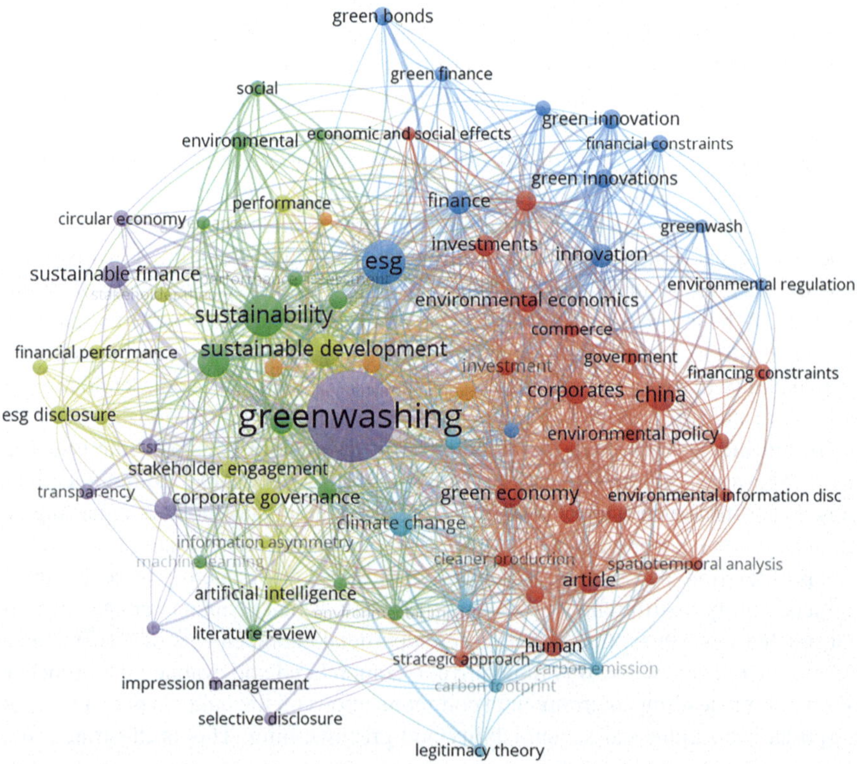

Fig. 4.6 Co-occurrence analysis. Source: Scopus/VOSviewer

ESG. Organizations are considerably more inclined to meet their commitments when they publicly disclose their environmental, social, and governance (ESG) metrics. Transparency facilitates accountability, which in turn encourages corporations to accomplish their ESG objectives.

Moreover, the transparency inherent in ESG reporting has the potential to expose instances of greenwashing by illuminating a company's authentic practices. The disclosure of a company's ESG activities allows stakeholders to assess the credibility of their claims. Moreover, transparency may function as a criterion for assessing the authenticity of a company's Environmental, Social, and Governance (ESG) initiatives, distinguishing between genuine efforts and mere greenwashing. A corporation may assert its commitment to investing in renewable energy; however, an accurate report would reveal the true magnitude of this investment. This would indicate whether the organization is solely procuring renewable energy credits while primarily depending on fossil fuels to meet its energy requirements. This level of transparency possesses the potential to uncover occurrences of greenwashing and to ensure that companies are held accountable for their Environmental, Social, and Governance (ESG) assertions (Hussain et al., 2023; Cahan et al., 2016; Nasrullah, 2011; Ma & Ahmad, 2024).

Moreover, transparency not only reveals instances of greenwashing but also plays a crucial role in substantially mitigating its occurrence. Organizations are less inclined to participate in greenwashing when they recognize that they will be held accountable for their Environmental, Social, and Governance (ESG) assertions. Moreover, transparency has the potential to impact investors' decision-making. Investors are progressively incorporating ESG factors into their investment decisions. Investors who prioritize sustainability are more likely to allocate their resources to organizations that demonstrate transparency in their Environmental, Social, and Governance (ESG) initiatives. This has the potential to instigate a constructive cycle that encourages organizations to improve their transparency and sustainability, consequently reducing the incidence of greenwashing. In conclusion, the mitigation of greenwashing in the context of Environmental, Social, and Governance (ESG) considerations is fundamentally reliant on transparency. This reveals the strategies employed for greenwashing and ensures that corporations are held accountable for their assertions regarding Environmental, Social, and Governance (ESG) criteria. Moreover, it assumes a pivotal role in the fight against greenwashing by advocating for authentic sustainability initiatives and shaping investor decisions (Karagiannis et al., 2022; Liu et al., 2020).

This co-occurrence study examines the phenomenon of greenwashing within the realm of sustainable finance, clarifying its effects on bonds and loans, the consequences it engenders, and the function of legal entities in fostering transparency and ethical practices within the industry. Greenwashing transpires when corporations or investment vehicles inaccurately assert their compliance with sustainability and Environmental, Social, and Governance (ESG) principles. The perception of a company's authentic dedication to sustainability may be compromised by these deceptive assertions, which can manifest in various forms and attract investors who prioritize social responsibility. In recent years, there has been a substantial increase in investor demand for green bonds and loans, driven by the escalating concerns related to environmental, social, and governance (ESG) factors (He et al., 2023; Bokhari et al., 2023; Oware & Mallikarjunappa, 2021). A green bond is differentiated from a conventional bond by its unique designation, which indicates a commitment to utilizing the funds raised solely for the financing or refinancing of projects, assets, or business activities that adhere to environmentally sustainable criteria. A green loan is designed to offer financial support for projects that meet established environmental criteria, akin to a green bond. The differentiation is evident in the scale and characteristics of the operation. Green loans are generally conducted in a private manner and are of a smaller magnitude, whereas green bonds typically encompass larger amounts, entail elevated transaction costs, and may be either listed on an exchange or issued privately.

Regrettably, greenwashing is a prevalent phenomenon within the global green-bond market. Greenwashing not only hinders progress in the fight against climate change but also raises significant concerns regarding financial stability. The divestment from and reassessment of a company's green bonds may be necessitated by the revelation of its misleading greenwashing practices, as investors increasingly incorporate environmental and sustainability considerations into their investment strategies. The identification of a substantial number of green-bond issuers as perpetrators of greenwashing may lead to a considerable exodus of capital, potentially intensifying price volatility within the green-bond market. Furthermore, the cascading effect may be broadened to encompass additional interrelated sectors, including green index funds; Environmental, Social, and Governance (ESG) or Socially Responsible Investment (SRI) funds, as well as green equities (Zournatzidou et al. 2024, 2024, 2024, 2024, 2024, 2024; Tzitiridou-Chatzopoulou et al. 2024; Russell et al., 2024).

4.5 Discussion

By the year 2025, the reporting of Environmental, Social, and Governance (ESG) themes is rapidly becoming the standard practice within the business sector. Prominent corporations are committing to attain net-zero emissions, while governmental entities are developing regulations and legislation to support this transition. Investors are acquiring a distinctive understanding of the environmental ramifications associated with corporate activities as a result of this development. ESG reporting is increasingly resembling financial reporting, as the most recent standards encompass both qualitative and quantitative disclosure obligations. Nevertheless, financial results are meticulously monitored and administered. Currently, ESG reporting does not exhibit the same level of rigor as financial statement reporting, which adheres to established standards and implements stringent controls to ensure consistency, accuracy, and comparability of data and disclosure. Consequently, a discrepancy may arise between a company's operations and its declarations. This discrepancy, which may occasionally be unintentional, has a detrimental impact on investors, consumers, employees, and all individuals who depend on this information for informed decision-making. This conduct is commonly referred to as "greenwashing," particularly in the context of sustainability.

Greenwashing is a tactic employed by corporations to portray themselves as more environmentally sustainable than they genuinely are. Greenwashing, which encompasses the dissemination of deceptive information regarding a product's sustainability or the erroneous classification of a fund as "green" when it does not meet such criteria, can have profound repercussions and erode public trust. It is imperative to acknowledge that greenwashing is not a fixed concept; rather, it exists along a continuum, extending from blatant deception to hopeful conviction. A corporation faces numerous hazards as a result of engaging in greenwashing practices. The initial impact of the quality of ESG data on disclosures is inconsistent due to variations in its accuracy and completeness. Product suppliers encounter considerable difficulties in acquiring the requisite data to guarantee adherence to sustainability standards. This is especially crucial for companies that underwrite policies in situations where there is uncertainty regarding the available or requisite data points or the criteria that constitute an acceptable ESG profile.

Furthermore, deficiencies in regulatory frameworks and insufficient standards significantly facilitate the continuation of greenwashing practices. The lack of obligatory legislation and adequate standards facilitates the continuation of greenwashing practices in the absence of consistent accountability mechanisms. Furthermore, the lack of explicit legislation exacerbates the issue by creating avenues for inadvertent greenwashing. In the absence of clearly delineated standards, companies may inadvertently adopt practices that misrepresent their environmental impact to consumers. Moreover, organizations must acknowledge the significant reputational risks associated with greenwashing. Mere rumors or suspicions of greenwashing may significantly damage a company's reputation. Consumer tolerance for greenwashing has significantly diminished, and the rapid spread of knowledge via social media, coupled with customers' ability to boycott brands, offers substantial risks to firms that make claims not substantiated by genuine actions. The potential ramifications and deterioration of consumer trust may have lasting implications on a company's reputation.

However, the current study underscored supplementary challenges for organizations in relation to greenwashing and its impact on ESG reporting. The phenomenon of ESG greenwashing presents considerable risks concerning governance. To promote transparency and avert misleading claims, it is imperative to implement comprehensive governance frameworks. Directors are obligated to guarantee that all environmental claims are precise, thoroughly documented, and aligned with the organization's comprehensive strategy and business model. This necessitates a commitment to transparency, meticulous documentation, and rigorous supervision. As a result, directors must ensure that their environmental assertions are accurate and substantiated when corporations address the increasing demand for sustainable practices. Greenwashing not only presents considerable risks but also diminishes the credibility of authentic environmental initiatives. Directors can safeguard their organizations from the perils of greenwashing and advance a more sustainable future by adhering to rigorous governance standards and fostering transparency.

The purpose of sustainability/ESG communication is to support companies in their long-term strategic endeavors, identify and mitigate substantial risks, and recognize potential growth opportunities. Despite a reduction in the implementation of anti-ESG measures in 2024, their deterrent impact remains evident, and considerable pressure from stakeholders continues to necessitate attention to social issues. A more integrated approach may be achieved by acquiring more exhaustive ESG data for specific topics, which will enable more informed decision-making, improved public policy outcomes, and enhanced brand messaging. Nonetheless, it is essential to assess the prospective antagonism of critics towards these objectives. Companies should be prepared to provide responses that emphasize the business rationale behind their operations, demonstrating their alignment with long-term value development and risk management.

This bibliometric investigation revealed the influence of greenwashing on loans. Greenwashing refers to the promotion of items as environmentally friendly or sustainable despite their failure to meet such criteria. In such cases, loans may not really provide significant environmental advantages, or may even promote ecologically harmful practices. Greenwashing may occur under the following circumstances: (i) when a loan is marketed as financing environmentally sustainable projects, yet the funds are not exclusively designated for such initiatives; (ii) when lenders fail to adequately evaluate the environmental consequences or sustainability credentials of the projects or activities financed by the loan; or (iii) when they employ vague or ambiguous language to describe the environmental benefits of a loan without providing a clear account of its actual ecological impact (Sun et al., 2024; Zhao et al., 2024). Thus, lenders may reduce the incidence of greenwashing in loans and provide borrowers with authentic possibilities to finance ecologically friendly projects by adopting effective strategies. This approach promotes responsibility, credibility, and openness in the financial sector's efforts to enhance its sustainability.

The results of this study may improve the current business environment and scholarly discussion. The literature emphasized a critical aspect of greenwashing: the understanding of ESG finance, an investment strategy that integrates environmental, social, and governance concerns. These factors are assessed in conjunction with conventional financial metrics to ascertain a company's social impact and sustainability. The primary objective of ESG finance is to ensure that investments are economically advantageous, while conforming to ethical, sustainable, and socially responsible criteria. The rising consumer demand for sustainable and ethical investments has resulted in an augmented need for ESG standards and regulations to effectively eliminate practices such as greenwashing. The regulation mandates institutions to provide comprehensive reports on governance, risk management, strategy, scenario analysis, and metrics. Businesses may use the findings for pragmatic insights, augmenting transparency, which is crucial for effectively combating greenwashing. Transparent reporting promotes accountability, nurtures confidence, and guarantees that sustainability claims correspond to actual action. Attaining transparency in reporting may be difficult without rigorous regulations and industry norms.

References

Bokhari, S. A. A., Ali, M., Albort-Morant, G., et al. (2023). Bridging the gap: The moderating roles of institutional quality and compliance on the link between CSR and financial performance. *Journal of Cleaner Production, 404*. https://doi.org/10.1016/j.jclepro.2023.136897

Cahan, S. F., De Villiers, C., Jeter, D. C., et al. (2016). Are CSR disclosures value relevant? Cross-country evidence. *European Accounting Review, 25*, 579–611. https://doi.org/10.1080/09638180.2015.1064009

Chueca Vergara, C., Ferruz Agudo, L., Rambaud, C., et al. (2021). Fintech and sustainability: Do they affect each other? *Sustainability, 13*, 7012. https://doi.org/10.3390/SU13137012

Demers, E., Hendrikse, J., Joos, P., & Lev, B. (2021). ESG did not immunize stocks during the COVID-19 crisis, but investments in intangible assets did. *Journal of Business Finance & Accounting, 48*, 433–462. https://doi.org/10.1111/JBFA.12523

Elshabshiri, A., Ghanim, A., Hussien, A., et al. (2025). Integration of building information modeling and digital twins in the operation and maintenance of a building lifecycle: A bibliometric analysis review. *Journal of Building Engineering, 99*. https://doi.org/10.1016/j.jobe.2024.111541

Galletta, S., Mazzù, S., Naciti, V., & Paltrinieri, A. (2024). A PRISMA systematic review of greenwashing in the banking industry: A call for action. *Research in International Business and Finance, 69*. https://doi.org/10.1016/J.RIBAF.2024.102262

He, F., Liu, G., Hao, J., & Li, Y. (2023). CSR performance and firm idiosyncratic risk in a data-rich environment: The role of retail investor attention. *Journal of International Financial Markets, Institutions and Money*, 101877. https://doi.org/10.1016/J.INTFIN.2023.101877

Huang, Y., Xiong, N., & Liu, C. K. (2024). Renewable energy technology innovation and ESG greenwashing: Evidence from supervised machine learning methods using patent text. *Journal of Environmental Management, 370*. https://doi.org/10.1016/j.jenvman.2024.122833

Hussain, T., Wang, D., & Benqian, L. (2023). Examining the role of responsible management, CSR, and TQM in enhancing renewable energy projects: An empirical analysis. *Acta Ecologica Sinica*. https://doi.org/10.1016/j.chnaes.2023.06.010

Jiang, C., Li, X., Xu, Q., & Liu, J. (2024). Does environmental protection tax impact corporate ESG greenwashing? A quasi-natural experiment in China. *Economic Analysis Policy, 84*, 774–786. https://doi.org/10.1016/J.EAP.2024.09.029

Karagiannis, I., Vouros, P., Sioutas, N., & Evangelinos, K. (2022). Mapping the maritime CSR agenda: A cross-sectoral materiality analysis of sustainability reporting. *Journal of Cleaner Production, 338*. https://doi.org/10.1016/j.jclepro.2021.130139

Khan, A. U., Ma, Z., Li, M., et al. (2023). *From traditional to emerging technologies in supporting smart libraries. A bibliometric and thematic approach from 2013 to 2022*. Library Hi Tech. ahead-of-print. https://doi.org/10.1108/LHT-07-2023-0280/FULL/XML

Lashitew, A. A. (2021). Corporate uptake of the sustainable development goals: Mere greenwashing or an advent of institutional change? *Journal of International Business Policy, 4*, 184–200. https://doi.org/10.1057/S42214-020-00092-4/FIGURES/1

Lee, M. T., & Raschke, R. L. (2023). Stakeholder legitimacy in firm greening and financial performance: What about greenwashing temptations?☆. *Journal of Business Research, 155*, 113393. https://doi.org/10.1016/J.JBUSRES.2022.113393

Li, M., & Chen, Q. (2024). Executive pay gap and corporate ESG greenwashing: Evidence from China. *International Review of Financial Analysis, 95*. https://doi.org/10.1016/J.IRFA.2024.103375

Li, Y., & Xiao, J. (2025). The effect of institutional investors' site visits on corporate greenwashing behavior. *International Review of Economics & Finance, 97*, 103818. https://doi.org/10.1016/J.IREF.2024.103818

Liu, Y., Lei, L., & Buttner, E. H. (2020). Establishing the boundary conditions for female board directors' influence on firm performance through CSR. *Journal of Business Research, 121*, 112–120. https://doi.org/10.1016/J.JBUSRES.2020.08.026

Ma, Y., & Ahmad, M. I. (2024). Do board characteristics impact greenwashing? Moderating role of CSR committee. *Heliyon, 10.* https://doi.org/10.1016/j.heliyon.2024.e38743

Migliorelli, M. (2021). What do we mean by sustainable finance? Assessing existing frameworks and policy risks. *Sustainability, 13*, 975. https://doi.org/10.3390/SU13020975

Moodaley, W., & Telukdarie, A. (2023). Greenwashing, sustainability reporting, and artificial intelligence: A systematic literature review. *Sustainability, 15*, 1481. https://doi.org/10.3390/SU15021481

Nasrullah, N. M. (2011). Business and sustainable development: The role of CSR as a catalyst. *Transnational Corporations Review, 3*, 96–105. https://doi.org/10.1080/19186444.2011.11658300

Oguntona, O. A. (2024). Evaluation of global research on greenwashing by scientometric indicators. *Sustainability and Toxicity of Building Materials*, 59–79. https://doi.org/10.1016/B978-0-323-98336-5.00004-2

Oware, K. M., & Mallikarjunappa, T. (2021). Financial performance and gender diversity: The moderating and mediating effect of CSR disclosure and expenditure of listed firms in India. *Vision: The Journal of Business Perspective, 27*, 243–255. https://doi.org/10.1177/09722629211001985

Page, M. J., McKenzie, J. E., Bossuyt, P. M., et al. (2021). The PRISMA 2020 statement: An updated guideline for reporting systematic reviews. *International Journal of Surgery, 88*. https://doi.org/10.1016/J.IJSU.2021.105906

Peng, D., & Kong, Q. (2024). Corporate green innovation under environmental regulation: The role of ESG ratings and greenwashing. *Energy Economics, 140.* https://doi.org/10.1016/j.eneco.2024.107971

Peng, Q., & Xie, Y. (2024). ESG greenwashing and corporate debt financing costs. *Finance Research Letters, 106012*. https://doi.org/10.1016/j.frl.2024.106012

Russell, A. C., Santucci, N. R., Tzitiridou-Chatzopoulou, M., et al. (2024). The potential impact of the gut microbiota on neonatal brain development and adverse health outcomes. *Children, 11*, 552. https://doi.org/10.3390/CHILDREN11050552

Smith, L. C., Mateos, A. C., Due, A. S., et al. (2024). Immersive virtual reality in the treatment of auditory hallucinations: A PRISMA scoping review. *Psychiatry Research, 334*. https://doi.org/10.1016/J.PSYCHRES.2024.115834

Sun, H., Bai, T., & Liu, Z. (2024). Perceived policy uncertainty and enterprise greenwashing risk. *Finance Research Letters, 69*. https://doi.org/10.1016/j.frl.2024.106148

Teti, E., Etro, L. L., & Pausini, L. (2024). Does greenwashing affect company's stock Price? Evidence from Europe. *International Review of Financial Analysis, 93*. https://doi.org/10.1016/J.IRFA.2024.103195

Torelli, R., Balluchi, F., & Lazzini, A. (2020). Greenwashing and environmental communication: Effects on stakeholders' perceptions. *Business Strategy and the Environment, 29*, 407–421. https://doi.org/10.1002/BSE.2373

Tzitiridou-Chatzopoulou, M., Zournatzidou, G., Tsakiridis, I., & Tsakalidis, C. (2024). Exploring the nexus between fertility rates and geopolitical risk with intelligence methods: A multifaceted analysis. *Healthcare, 12*, 2205. https://doi.org/10.3390/HEALTHCARE12222205

Uyar, A., Karaman, A. S., & Kilic, M. (2020). Is corporate social responsibility reporting a tool of signaling or greenwashing? Evidence from the worldwide logistics sector. *Journal of Cleaner Production, 253*, 119997. https://doi.org/10.1016/J.JCLEPRO.2020.119997

Wang, Z., & Sarkis, J. (2017). Corporate social responsibility governance, outcomes, and financial performance. *Journal of Cleaner Production, 162*, 1607–1616. https://doi.org/10.1016/J.JCLEPRO.2017.06.142

Wang, Z., Hsieh, T. S., & Sarkis, J. (2018). CSR performance and the readability of CSR reports: Too good to be true? *Corporate Social Responsibility and Environmental Management, 25*, 66–79. https://doi.org/10.1002/CSR.1440

Wang, Q., Ma, Z., Zhao, J., & Shu, G. (2024). State shareholding in privately-owned firms and greenwashing: Evidence from China. *Finance Research Letters, 62*. https://doi.org/10.1016/J.FRL.2024.105176

Wang, Y., Xing, C., & Zhang, L. (2024). Is greenwashing beneficial for corporate access to financing? Evidence from China. *Finance Research Letters, 65*. https://doi.org/10.1016/J.FRL.2024.105538

Wang, Y., Hu, F., & Wang, Y. (2024). Analyst coverage and greenwashing: Evidence from Chinese A-share listed corporations. *International Review of Economics and Finance, 94*. https://doi.org/10.1016/j.iref.2024.103423

Yu, E. P., Van Luu, B., & Chen, C. H. (2020). Greenwashing in environmental, social and governance disclosures. *Research in International Business and Finance, 52*, 101192. https://doi.org/10.1016/J.RIBAF.2020.101192

Zhang, D. (2022a). Are firms motivated to greenwash by financial constraints? Evidence from global firms' data. *Journal of International Financial Management & Accounting, 33*, 459–479. https://doi.org/10.1111/JIFM.12153

Zhang, D. (2022b). Green financial system regulation shock and greenwashing behaviors: Evidence from Chinese firms. *Energy Economics, 111*, 106064. https://doi.org/10.1016/J.ENECO.2022.106064

Zhao, W., Zhong, M., Liao, X., et al. (2022). Board network and CSR decoupling: Evidence from China. *Frontiers in Psychology, 13*, 815341. https://doi.org/10.3389/FPSYG.2022.815341/BIBTEX

Zhao, X., Huang, X., Liu, F., & Pan, L. (2024). Executive power discrepancy and corporate ESG greenwashing. *International Review of Financial Analysis, 96*. https://doi.org/10.1016/j.irfa.2024.103533

Zhou, K., Qu, Z., Liang, J., et al. (2024). Threat or shield: Environmental administrative penalties and corporate greenwashing. *Finance Research Letters, 61*. https://doi.org/10.1016/J.FRL.2024.105031

Zournatzidou, G., Ragazou, K., Sklavos, G., et al. (2024). (1AD) The influence of greenwashing methods on the behavior of environmentally conscious employees. In R. Konstantina, G. Alexandros, P. Christos, & S. Angeliki (Eds.), *Sustainability through green HRM and performance integration* (pp. 39–58). IGI Global.

Zournatzidou, G., Ragazou, K., Samantzis, C., et al. (2024). (1AD) Looking back and looking forward: Current and future practices in green human resources management (RM) and green innovation. In R. Konstantina, G. Alexandros, P. Christos, & S. Angeliki (Eds.), *Sustainability through green HRM and performance integration* (pp. 441–458). IGI Global.

Zournatzidou, G., Sklavos, G., Ragazou, K., & Sariannidis, N. (2024). (1AD) Building carbon-neutral green HRM resources through the use of green financing. In R. Konstantina, G. Alexandros, P. Christos, & S. Angeliki (Eds.), *Sustainability through green HRM and performance integration* (pp. 219–242). IGI Global.

Zournatzidou, G., Farazakis, D., Sklavos, G., et al. (2024). Sustainability, green financing, and corporate management. In R. Konstantina, G. Alexandros, P. Christos, & S. Angeliki (Eds.), *Sustainability through Green HRM and performance integration* (pp. 191–218). IGI Global.

5 Controversies Over Anti-competition and Anti-corruption: Assessing Anti-ESG Variables Using Entropy-Weighted TOPSIS Methodologies in the European Banking Sector

5.1 Introduction

The financial industry has a significant issue in the form of corruption regarding the safeguarding of human rights and the enforcement of laws. This issue has emerged as a significant worry across the financial industry, including financial markets, global finance, governmental money, business finance, and personal finance. Knack and Keefer (1997), Liu et al. (2023b), and Sun et al. (2023) demonstrate that corruption is a significant and enduring obstacle to economic progress and social development in any nation. This study was undertaken by experts in the field. Moreover, Basnyat et al. (2023) assert that the proliferation of organized criminal enterprises is intricately linked to the occurrence of corruption. Corruption within the financial ecosystem undermines competitiveness, hinders local and foreign investments, and diminishes economic growth (Balsalobre-Lorente et al., 2023; Priya & Sharma, 2023; Su & Xu, 2023). Moreover, corruption has a profoundly adverse effect on some nations, especially those characterized by diminished democratic governance. Corruption substantially diminishes both economic development and wealth distribution. The acts impede the operations and development of other firms that comply with legal standards. The prevalence of employment insecurity, distrust, and dissatisfaction among workers and employers is particularly concerning in the context of emerging economies (Al-Sadig, 2009; Basnyat et al., 2023; Cuervo-Cazurra, 2006). Finally, corruption within financial institutions not only diminishes the legitimacy of these entities and the market, both nationally and globally, but also threatens to undermine public trust in institutions, potentially leading to their downfall (Liu et al., 2023b; Lucey et al., 2023; Su & Xu, 2023; Sun et al., 2023).

Corruption in the financial industry often presents as money laundering, financial fraud, tax evasion, exchange restrictions, stock manipulation, and embezzlement. The additional manifestations of corruption are embezzlement and stock manipulation. Consequently, there has been a paucity of studies on the examination of

K. Ragazou et al., *Transparency and Corporate Washing in the Age of Sustainability*, Management for Professionals,
https://doi.org/10.1007/978-3-031-96821-1_5

anti-competitive behaviors and corruption within the banking industry. Moreover, no research has been published on the examination of anti-competitive disputes and corruption within the industry (Shakil, 2021). This report is the first investigation of anti-competition problems and their influence on the proliferation of corruption, bribery, and fraud inside the European banking sector.

Although anti-competitive and corrupt practices seem to be separate concepts, both lead to disproportionate market domination within the financial industry. This, therefore, results in inefficiencies and escalates expenses, thereby impacting the price and quality of financial goods and services. These two offenses often have synergistic consequences, leading to an inability to get optimal value for money. This is achieved by granting contracts or markets based on criteria that do not reflect fair competition and the contractor's proficiency. Moreover, bribery and corrupt activities are often associated with collusive and anti-competitive conduct. Examples of these practices include price manipulation, market division, and bid rigging. Consequently, several investigations include allegations of antitrust violations and corruption (Basnyat et al., 2023; Cuervo-Cazurra, 2006).

This research is the first study to examine the relationship between the three components of ESG (Environmental, Social, and Governance) and their impact on bank corruption, as shown by instances of anti-competitive events (Liu et al., 2023b). This research was conducted by J. Liu and associates. Our contribution aims to address a gap in the existing literature by answering the following research question: How can we identify the most and least corrupt financial institutions in Europe, and what characteristics should the market consider when developing an effective anti-corruption strategy? We have created two models using multiple-criteria decision analysis (MCDM) to assess the elements. The first model is the technique for order preference by similarity to ideal solution (TOPSIS) approach, whereas the subsequent model is the entropy-weighted TOPSIS method. Both concepts are based on the TOPSIS methodology.

The subsequent sections of the paper are organized as follows: The second portion presents a review of the existing literature, while the third section details the data and methodologies used in the inquiry. The fourth portion presents the empirical findings, while the fifth section analyzes the data and discusses its theoretical and management implications. Additionally, it proposes potential areas for future investigation. The article concludes with the seventh and last section—the conclusion.

5.2 Literature Review

5.2.1 The European Financial Industry's Emerging Form of Corruption: Anti-competition Controversies

Over the past few decades, a significant quantity of research has been conducted to examine the conceptualization and definition of corruption in both the public and private sectors. It is challenging to ascertain the precise meaning of the phrase, as it could be referring to a particular sector, a collection of distinct firms, or an interdisciplinary perspective. The United Nations has identified the reduction of corruption as a significant priority in the context of the Sustainable Development Goals (SDGs), as per Chu et al. (2023) and Roy et al. (2023). Corruption is a multifaceted and intricate issue that impacts not only governments but also public institutions and private enterprises. The World Bank and Transparency International's definition is highly regarded and quite persuasive. Corruption is defined by Transparency International (TI) as the "inappropriate utilization of power for personal gain," while the World Bank defines it as the "manipulation of a public role for personal financial advantage." The concept is the same in both definitions. The first form of corruption is corruption in the public environment, while the second type is corruption in the private sector. The issue of corruption has been exacerbated by the assistance of financial institutions and banks in the concealment of illicit funds, as per Amankwaah and Baidoo (2023).

There are numerous forms of corruption that may manifest within financial institutions. Embezzlement, abuse of money, fabrication of financial statements and documents, cyber fraud, extortion, bribery, suppression of facts, and violation of various laws and regulatory frameworks are but a few examples of these types of corruption. Violations of confidentiality, the misuse of privileged information for personal gain, and the transfer or reception of illicit currency are among the other criminal activities that may be observed in the financial sector. Additionally, a substantial portion of the corruption that infiltrates the financial system can be attributed to either the illicit activities of organizations or the unlawful actions of customers. One instance of this is the utilization of falsified documents and the attempt to pilfer user information in order to conduct online banking transactions.

However, the global financial crisis and the COVID-19 epidemic have sparked a heightened interest among policymakers and scholars in the competition between banks, with a particular emphasis on issues related to anti-competitive activities. The simultaneous occurrence of both of these events has served as an incentive for this attention. Corrupt practices are becoming increasingly common in the financial sector, as evidenced by the frequency of these incidents. The academic community is of the opinion that there is a prevalent belief that an increase in anti-competitive activity could potentially result in financial instability. Furthermore, there is concern regarding the heightened level of integration among financial institutions that has been the consequence of the global financial crisis and the pandemic humanitarian crisis. This may lead to the implementation of measures to prevent future instability and unjust competition. This is due to the ethical concerns that are associated with financial institutions that are perceived as being too large to fail.

The prevalent belief that European financial institutions engage in anti-competitive behaviors that are particularly severe is illustrated in Fig. 5.1(a–e). The objective of European banks is to establish a European financial system that is competent and capable of efficiently accommodating the requirements of European enterprises, investors, and citizens. It is imperative to provide European businesses, investors, and individuals with the necessary financial resources in order to thrive in a global economy that is exceedingly competitive. This holds true irrespective of the business's size, geographical location, or type. The establishment of a financial

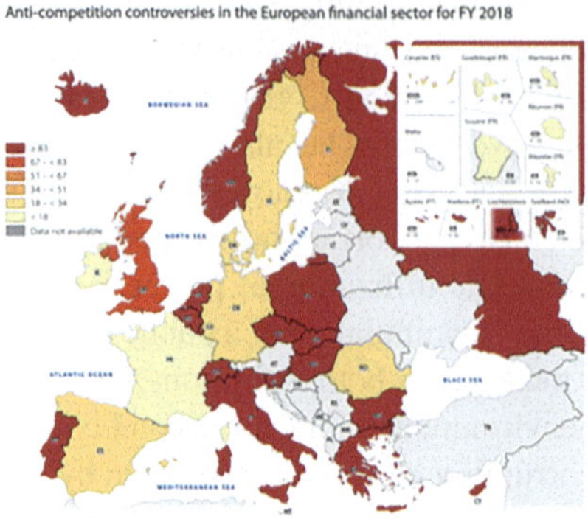

1.a

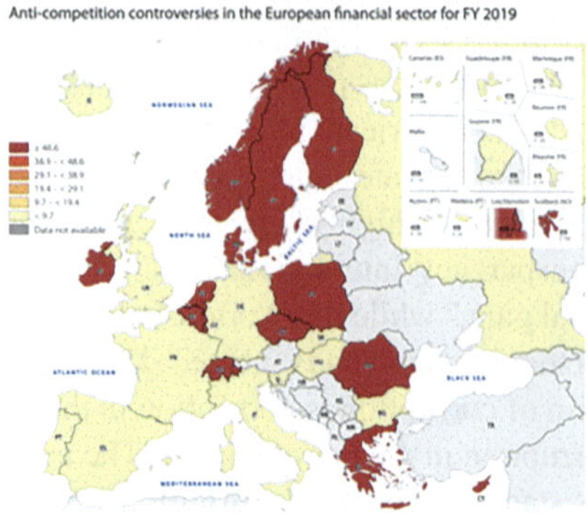

1.b

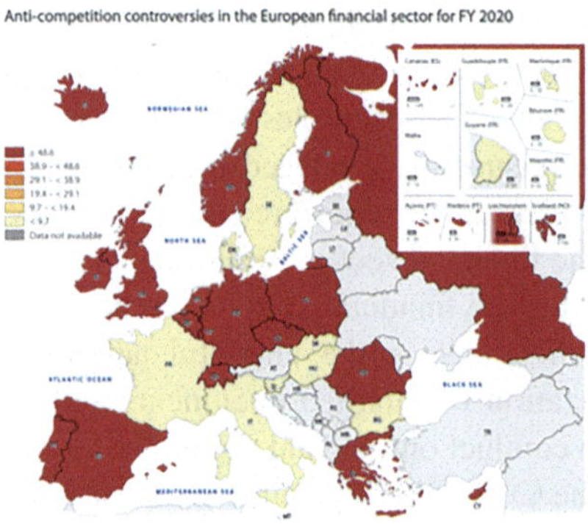

1.c

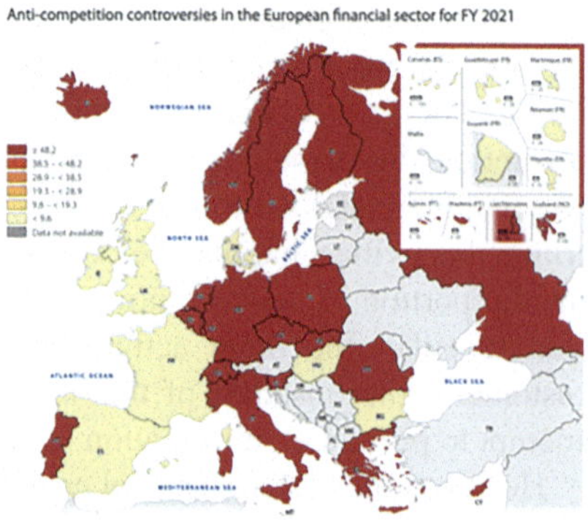

1.d

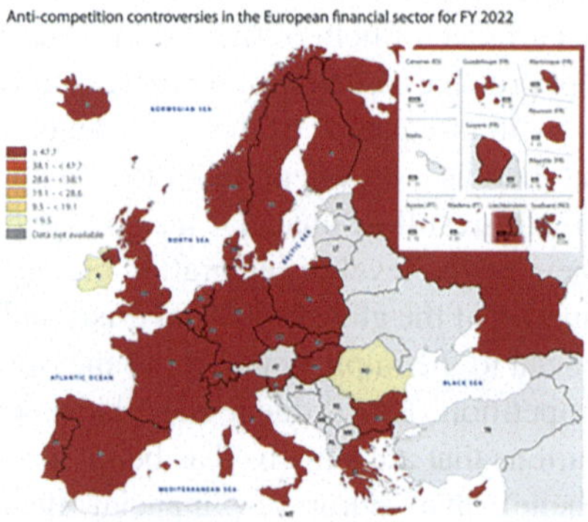

1.e

Fig. 5.1 (a–e) The European financial sector's spatial patterns of anti-competition controversies. Source: Authors' elaboration

sector with the objective of minimizing the creation of liabilities, excessive risks, and losses for taxpayers and customers has been the primary focus of European policymakers and the financial industry over the past ten years, despite the apparent simplicity of the matter. Unavoidably, the emphasis was shifted to prioritizing safety, stability, and investor protection over the efficiency or productivity of the financial sector as a result of the calamitous global financial crisis, which posed a threat to economic development. This transition was precipitated by the crisis's potential to undermine economic expansion. Fig. 5.1 (a–e) illustrate the extent of the anti-competitive conflicts that financial institutions have engaged in in each European nation (Cicchiello et al., 2023; DasGupta, 2022). Given these circumstances, it appears that Europe has not yet achieved its goal of establishing a financial system that is both equitable and transparent.

The position in the Scandinavian financial market is one of the most astonishing revelations regarding the extent of anti-competition concerns in the European financial sector. As stated by Latapí et al. (2023), the Scandinavian region has consistently been recognized as one of the most technologically advanced regions in the world. The fact that Scandinavian banks are currently embroiled in a significant controversy, which encompasses money laundering and anti-competitive activities, is truly remarkable. These financial companies are being accused of facilitating the transfer of massive quantities of illegal funds, totaling hundreds of billions of dollars, to offshore tax havens on behalf of Russian billionaires, corrupt politicians, and leaders of organized crime (Lindman et al., 2020; Tchapchet Tchouto, 2023) (Lindman et al., 2020; Tchapchet Tchouto, 2023). For instance, Danske Bank, Denmark's most significant financial institution, was initially implicated in illicit activities at its Estonian branch. Danske Bank only became aware of the issues in 2017 when they were brought to the attention of the Danish media. The extent of the transgression was not disclosed until September 2018, when Danske Bank disclosed the results of an inquiry conducted by an independent law firm. During the same period that the bank acknowledged its "blatant failure to fulfill its obligation," Thomas Borgen, the chief executive officer, resigned in response to the increasing volume of criticism.

Additionally, the Greek banking sector serves as an exceptional illustration, as it is distinguished by its elevated anti-competition ratings. The Hellenic Competition Commission (HCC) is responsible for the enforcement of antitrust legislation that has been established by both Greece and the European Union within Greek jurisdiction. To be more precise, this includes the implementation of Articles 1 and 2 of the Competition Act, as well as Articles 101 and 102 of the Treaty on the Functioning of the European Union (TFEU). Any agreements, decisions, or practices that impede competition between enterprises are prohibited by Article 1 of the Competition Act and Article 101 of the Treaty on the Functioning of the European Union. Both Article 2 of the Competition Act and Article 102 of the Treaty on the Functioning of the European Union prohibit the misuse of a dominant market position. These anti-competitive actions are closely linked to the financial institutions of Greece, which are situated within the country. Commercial institutions are frequently accused of malfeasance. However, a number of the world's most prominent financial institutions have utilized low-interest loans as an incentive to acquire profitable services that were previously performed by investment banks. This type of accusation is not uncommon in the financial sector of Greece and other countries. For an extended period, the combination of loan decisions with other banking activities was prohibited. Any astute financier would vigorously deny that their financial institution was involved in such behavior from the outset, with the assistance of their legal counsel. Although the concept of linking grants and other activities may appear straightforward, the actual presentation of this connection is occasionally absent, and the constraints can be easily evaded. For example, a bank is permitted to conduct a thorough examination of all transactions that it has conducted with a client before determining whether or not to continue providing financial assistance to that client. In the interim, consumers are permitted to express their willingness to establish a business partnership with a financial institution, provided that they are able to obtain loans at rates that are within their financial reach. Conversely, the Romanian financial sector serves as an illustration of equitable competition within the financial institution sector. According to the Competition Council, Romanian law prohibits any agreements or coordinated activities among financial institutions that obstruct, limit, or distort competition in the Romanian financial sector. This prohibition pertains to financial institution collaboration. This encompasses the establishment of prices for the purpose of purchasing or selling, as well as any other terms that are related to trading.

5.2.2 ESG Controversies: The European Financial Sector's Impact on Corruption in ESG Standards

Notwithstanding the efforts of national and international regulatory entities, including central banks, agencies, and standards set out by the Organization for Economic Co-operation and Development (OECD) and the International Monetary Fund (IMF), corruption persists. Corruption is the fundamental cause of the proliferation of poverty, greed, unemployment, fragile institutions, and inadequate law enforcement. The Global Witness Organization (GWO) asserts that corrupt individuals, such as businesses, government officials, warlords, and criminals, need a financial institution to consistently conceal and legitimize their illicitly obtained funds. Consequently, banks and other financial organizations play a crucial role in the concealment of illicit funds. The inadequate corporate governance of banks, the incompetence of bank management, and the participation of businesses in unethical practices are all reasons attributed to the bankruptcy of financial institutions in the past, such as Bear Stearns and Lehman Brothers. Regulatory authorities and university lecturers have conducted investigations and research on bank corruption due to scandals affecting financial firms.

Policymakers, practitioners, stakeholders, and researchers have increasingly recognized the growing focus on corruption within the financial sector. This emphasis is primarily on the economic and financial elements that contribute to the increase in the occurrences of corruption, fraud, and bribery. Compensation is not allocated to entities involved in the development and distribution of environmental, social, and governance (ESG) indicators. ESG is an abbreviation for environmental, social, and governance factors. Rather than focusing just on economic and financial factors, ESG encompasses all three aspects. In the realm of corporate governance and financial investments, these factors are acquiring heightened importance (Agnese et al., 2023; Cernohorska, 2015; Khan et al., 2022).

The ESG controversies rating evaluates a company's susceptibility to environmental, social, and governance challenges and adverse events as reported in the worldwide media (Agnese et al., 2023; Anita et al., 2023; Bang et al., 2023). This grade is derived from the company's perception in the media. Recent conversations among many authors have focused on the role of the ESG debate concerning non-financial enterprises. Li et al. (2019) examine a dataset of 2613 enterprises from 39 different countries, covering the period from 2009 to 2013. Their assertion is that when companies face a conflict, they endeavor to mitigate the potential harm to their credibility, reputation, and stakeholder relationships by engaging in symbolic actions classified as corporate social responsibility (CSR). Aouadi and Marsat (2018) illustrate a correlation between environmental, social, and governance (ESG) concerns and heightened company market value. They studied a large dataset including over 4000 enterprises from 58 different countries to achieve this objective. DasGupta (2022) examines a worldwide dataset including data from 27 nations, covering the period from 2010 to 2019, as part of a comparable research program. The author's theory posits that companies experiencing financial underperformance and significant environmental, social, and governance (ESG) criticism are more likely to attain a commendable level of ESG performance. Marsat et al. (2022) concentrate on environmental concerns as their principal subject of inquiry. The authors show, using a dataset of 233 global environmental conflicts from 2010 to 2016, that environmental performance significantly bolsters the resilience of enterprises facing environmental challenges.

In addressing corruption within the financial sector, policymakers, practitioners, stakeholders, and researchers tend to focus mostly on the economic and financial factors that may lead to an increase in bribery, fraud, and corruption. Environmental, social, and governance (ESG) conflicts often neglect the factors that lead to their emergence and occurrence. ESG encompasses the three components of environmental, social, and governance factors, extending beyond just economic and financial issues. ESG is an abbreviation for environmental, social, and governance. In the realms of corporate governance and financial investments, these attributes are becoming more significant. The ESG controversy rating evaluates a firm's vulnerability to environmental, social, and governance issues, as well as negative incidents reported in the global media. Recent research by several authors has examined the impact of environmental, social, and governance (ESG) issues on non-financial organizations; however, none have explicitly addressed the role of ESG factors, particularly those that contribute to challenges and foster corruption within the financial sector.

This research is the first study to examine environmental, social, and governance (ESG) discussions within the banking industry. This research aims to assess the extent to which the dynamics of environmental, social, and governance (ESG) conflicts contribute to anti-competitive practices, corruption, and illegal activities inside European financial institutions. We will do an analysis of a panel dataset including

Table 5.1 Criteria and their computation units

Criteria	Description	Measurement
C_1	Policy fair competition score	Does the corporation explicitly state in its code of conduct that it endeavors to be a fair competitor? Encompasses the adherence to the patents, copyrights, and intellectual properties of other companies, as well as the avoidance of anti-competitive behavior, price fixing, and other monopolistic practices Extract data from the code of conduct section in each given report.
C2	Policy bribery and corruption score	Does the company explicitly state in its code of conduct that it endeavors to prevent bribery and corruption in all of its operations? The company has implemented a policy in its code of conduct to address and prevent bribery and corruption in its operations. Any report should take into account the information provided in the code of conduct section. Legal compliance data is not taken into consideration. This policy covers activities such as inappropriate or improper payments, special favors, extortion, or kickbacks
C3	Bribery, corruption, and fraud controversies score	Is the corporation now facing public scrutiny due to any controversies related to bribery, corruption, political donations, unethical lobbying, money laundering, unauthorized imports, or tax evasion?
C4	ESG controversies score	The ESG controversies category score quantifies a company's level of involvement in environmental, social, and governance disputes and unfavorable occurrences as reported in international media
C5	Wages and working condition controversies score	Is the company currently facing public scrutiny owing to controversies involving its employees, contractors, or suppliers, specifically related to wage disputes, layoffs, or working conditions?
C6	Consumer complaints controversies score	Is the company receiving significant media attention due to consumer complaints or dissatisfaction specifically related to its products or services?
C7	Product quality controversies score	Is the corporation currently facing public scrutiny due to an issue about the quality and responsibility of its products or services?
C8	Accounting controversies score	Is the corporation currently facing public scrutiny due to a controversy related to aggressive or non-transparent accounting practices?
C9	Responsible marketing controversies score	Is the corporation currently facing public scrutiny due to an issue related to its marketing methods, including the over-promotion of unhealthy food to susceptible consumers?
C10	Executive compensation controversies score	Is the corporation attracting public attention due to a scandal around excessive compensation for high-ranking executives or board members?
C11	Insider dealings controversies score	Is the firm currently facing public scrutiny due to a problem involving insider trading and other forms of share price manipulation?

(continued)

Table 5.1 (continued)

Criteria	Description	Measurement
C12	Policy business ethics score	Does the organization explicitly state in its code of conduct that it is committed to upholding the utmost standards of overall business ethics? Details regarding adherence to universal principles of corporate ethics or integrity Data derived from the part pertaining to the code of conduct
C13	Environmental controversies	Is the corporation currently receiving significant media attention due to an issue around the environmental impact of its operations on natural resources or local communities?
C14	Emissions score	The emission category score quantifies a company's dedication and efficacy in mitigating environmental emissions during its manufacturing and operational activities

344 European financial institutions, covering the years 2018 to 2022. We included 14 criteria pertaining to environmental, social, and governance (ESG) disputes, derived from the three sub-pillars of ESG (Table 5.1), to examine the impact of the selected components of ESG controversies. Our company rigorously complies with the classifications and subdivisions established by the reference data supplier.

5.3 Materials and Methods

The TOPSIS method is a structured methodology used for decision-making using several criteria within a specified framework. The entropy-weighted TOPSIS model is a hybrid approach that combines the entropy technique with the TOPSIS method. Kaur et al. (2023). The primary function of this system is to ascertain the weight of each evaluation indication using the entropy-weighted methodology. It then utilizes a strategy to approach the ideal answer for determining the ranking of the evaluation items. The fundamental objective of the entropy-weighted TOPSIS method is to identify the ideal solution, whereby all attribute values attain their optimum (or suboptimal) levels within the alternative framework (Dwivedi & Sharma, 2023). The evaluation object is considered optimum if it is nearest to the best solution and furthest from the worst option, based on the comparative distances to each. If it does not meet these characteristics, it is deemed suboptimal. The notion of the entropy-weighted TOPSIS method effectively utilizes information from the original dataset without placing specific constraints on the sample size. It offers the benefits of little information loss and adaptable functionality (Atenidegbe & Mogaji, 2023; Zare Banadkouki, 2023).

Prior to presenting the approaches, we make the initial assumption that there are n alternatives $A = \{A_1, A_2, A_3, \ldots, A_n\}$, and m criteria $M = \{M_1, M_2, M_3, \ldots, M_m\}$, where, $i \in A, j \in M, i = \{1, 2, 3, \ldots, n\}, j = \{1, 2, 3, \ldots, m\}$.

The matrix $X' = (x'_{ij})_{nxm}$ in Eq. (5.1) is a decision matrix of $n \times m$. The weights of the criteria M can be represented by the weight vector $W = \{w_1, w_2, w_3, \ldots, w_m\}$, which satisfies $\sum_{j=1}^{m} w_j = 1$.

$$X' = \begin{bmatrix} x'_{11} & x'_{12} & x'_{1m} \\ x'_{21} & \ldots & x'_{2m} \\ x'_{n1} & x'_{n2} & x'_{nm} \end{bmatrix} \tag{5.1}$$

5.3.1 Entropy-Weighted Method

Shannon's entropy-weighted method can ascertain the weight based on the level of data dispersion (Liu et al., 2023a; Saraiva, 2023). Initially, we employ the Min–Max technique to normalize the $n \times m$ original decision matrix $\{x'_{ij}\}$. Subsequently, we shift the standardized formula to the right by 0.001 units to simplify subsequent logarithmic computations.

$$x_{ij} = \frac{x'_{ij} - \min(x'_j)}{\max(x'_j) - \min(x'_j)} + 0.001, \text{where } i = 1,2,3,\ldots,n, \text{and } j = 1,2,3,\ldots,m. \tag{5.2}$$

The entropy value, which is denoted as e_j has been computed by using Eq. (5.4). Entropy quantifies the level of data dispersion. As the data becomes more spread out, the entropy number decreases, indicating that the data contains a greater amount of information. As the data becomes more concentrated, the entropy value increases, indicating a decrease in the amount of information contained in the data.

$$r_{ij} = \frac{x_{ij}}{\sum_{i=1}^{n} x_{ij}}, i = 1,2,3,\ldots,n, \text{and } j = 1,2,3,\ldots,m. \tag{5.3}$$

$$e_j = -\frac{1}{\ln n} \sum_{i=1}^{n} r_{ij}, i = 1,2,3,\ldots,n \text{ and } j = 1,2,3,\ldots,m. \tag{5.4}$$

The weight w_j is calculated by using Eq. (5.5).

$$w_j = \frac{1 - e_j}{\sum_{j=1}^{m} (1 - e_j)} \tag{5.5}$$

5.3.2 TOPSIS Model

The TOPSIS model, which was developed by Hwang and Yoon, is a method for assessing the alternatives by determining how closely they are related to the ideal solutions (Hwang & Yoon, 1981). The Euclidean distance between each goal choice

Table 5.2 The weight of each criterion for FYs 2022–2018 based on the entropy-weighted TOPSIS method

FY 2022														
Criterion	**C1**	**C2**	**C3**	**C4**	**C5**	**C6**	**C7**	**C8**	**C9**	**C10**	**C11**	**C12**	**C13**	**C14**
ej	−0.8471	−0.9498	−0.9827	−0.9926	−0.9985	−0.9865	−0.9897	−0.9985	−0.9955	−0.9975	−0.9995	−0.9170	−0.3332	−0.9661
$d = 1 - ej$	0.1529	0.0502	0.0173	0.0074	0.0015	0.0135	0.0103	0.0015	0.0045	0.0025	0.0005	0.0830	0.6668	0.0339
Wj	0.0267	0.0088	0.0030	0.0013	0.0003	0.0024	0.0018	0.0003	0.0008	0.0004	0.0001	0.0145	0.1165	0.0059
FY 2021														
Criterion	**C1**	**C2**	**C3**	**C4**	**C5**	**C6**	**C7**	**C8**	**C9**	**C10**	**C11**	**C12**	**C13**	**C14**
ej	−0.8384	−0.9366	−0.9771	−0.9911	−0.9955	−0.9892	−0.9934	−0.9944	−0.9955	−0.9960	−0.9949	−0.9095	−0.4255	−0.9567
$d = 1 - ej$	0.1616	0.0634	0.0229	0.0089	0.0045	0.0108	0.0066	0.0056	0.0045	0.0040	0.0051	0.0905	0.5745	0.0433
Wj	0.0282	0.0111	0.0040	0.0015	0.0008	0.0019	0.0012	0.0010	0.0008	0.0007	0.0009	0.0158	0.1004	0.0076
FY 2020														
Criterion	**C1**	**C2**	**C3**	**C4**	**C5**	**C6**	**C7**	**C8**	**C9**	**C10**	**C11**	**C12**	**C13**	**C14**
ej	−0.8164	−0.9158	−0.9604	−0.9762	−0.9838	−0.9731	−0.9811	−0.9833	−0.9838	−0.9844	−0.9838	−0.8925	−0.5823	−0.9412
$d = 1 - ej$	0.1836	0.0842	0.0396	0.0238	0.0162	0.0269	0.0189	0.0167	0.0162	0.0156	0.0162	0.1075	0.4177	0.0588
Wj	0.0321	0.0147	0.0069	0.0042	0.0028	0.0047	0.0033	0.0029	0.0028	0.0027	0.0028	0.0188	0.0730	0.0103
FY 2019														
Criterion	**C1**	**C2**	**C3**	**C4**	**C5**	**C6**	**C7**	**C8**	**C9**	**C10**	**C11**	**C12**	**C13**	**C14**
ej	−0.7930	−0.8913	−0.9341	−0.9542	−0.9629	−0.9540	−0.9573	−0.9623	−0.9592	−0.9598	−0.9623	−0.8718	−0.7147	−0.9154
$d = 1 - ej$	0.2070	0.1087	0.0659	0.0458	0.0371	0.0460	0.0427	0.0377	0.0408	0.0402	0.0377	0.1282	0.2853	0.0846
Wj	0.0362	0.0190	0.0115	0.0080	0.0065	0.0080	0.0075	0.0066	0.0071	0.0070	0.0066	0.0224	0.0499	0.0148
FY 2018														
Criterion	**C1**	**C2**	**C3**	**C4**	**C5**	**C6**	**C7**	**C8**	**C9**	**C10**	**C11**	**C12**	**C13**	**C14**
ej	−0.7773	−0.8639	−0.9227	−0.9330	−0.9405	−0.9311	−0.9326	−0.9412	−0.9369	−0.9405	−0.9398	−0.8519	−0.7885	−0.8793
$d = 1 - ej$	0.2227	0.1361	0.0773	0.0670	0.0595	0.0689	0.0674	0.0588	0.0631	0.0595	0.0602	0.1481	0.2115	0.1207
Wj	0.0389	0.0238	0.0135	0.0117	0.0104	0.0120	0.0118	0.0103	0.0110	0.0104	0.0105	0.0259	0.0370	0.0211

among European financial institutions from 2019 to 2022 through the application of the entropy-weighted TOPSIS technique. Financial institutions are of paramount importance in the process of facilitating the transition to a more ecologically balanced society. They have the capacity to influence markets in order to achieve specific objectives by limiting financing or altering the value of financial instruments. For example, they may prohibit activities that are harmful to the environment (also referred to as "brown" activities) and advertise activities that are environmentally friendly (also known as "green" activities). Concurrently, banks will assume a significant role in mitigating the consequences of these risks, as bank borrowers are increasingly confronted with the challenges posed by environmental degradation and climate change. In order to establish a robust regulatory framework for the management of C&E risks, regulatory authorities in Europe have implemented regulations that mandate financial institutions to incorporate C&E-specific objectives and key performance indicators (KPIs) into their business strategies and risk management. Major financial institutions have formulated their own sustainability commitments, particularly in relation to environmental concerns, in order to satisfy the increasing expectations of regulatory agencies. This has highlighted the necessity of identifying and implementing pertinent key performance indicators (KPIs) and key risk indicators (KRIs) in addition to establishing and implementing strategies that are associated with ethics and compliance.

In contrast, financial organizations do not implement the aforementioned techniques to a significant extent. Financial institutions worldwide, particularly in Europe, are significantly contributing to the environmental degradation by making substantial investments in the oil and gas sector. Businesses that are involved in the fossil fuel industry may capitalize on the loans, insurance, and other financial instruments that financial institutions explicitly provide. In order to support the operations of those who mine coal and consume oil, they do this, which ultimately leads to the emission of significant quantities of greenhouse gases into our environment. Additionally, financial institutions possess the capacity to provide funding for specific initiatives, such as the expansion of a coal mine or the construction of a new oil platform. Without sufficient financial resources, this type of undertaking would be exceedingly challenging to execute. Nevertheless, their influence on the climate surpasses these physical limitations. The financial sector primarily provides funding to businesses that are accountable for substantial environmental impacts. The fabrication of weapons and animal agriculture are among the activities that these enterprises engage in. British banks also support businesses that are responsible for some of the most severe environmental violations, including the extensive devastation of the Amazon rainforest. The fossil fuel sector is currently receiving financial support from a substantial number of high-street institutions in the United Kingdom. The fossil fuel sector has received billions of dollars in funding from a significant number of financial institutions in the United Kingdom as a consequence of the Paris Agreement's passage in 2015, which established the objective of limiting global warming to well below 2 degrees Celsius. To be more precise, the Banking on Climate Chaos research encompasses numerous prominent high-street banks in the United Kingdom as part of the top sixty institutions that offer financial support to the fossil fuel sector. The investigation encompasses eight distinct banking organizations. Instructions for ethical current accounts, savings accounts, and mortgages, among other accounts, address these groups. Thus, despite the fact that an increasing number of European financial institutions are emphasizing their dedication to environmental sustainability, the assurances they make frequently fail to withstand close examination. Despite the fact that certain fossil fuels are prohibited by certain organizations, these regulations frequently contain loopholes that suggest corruption and anti-competitive behavior.

In the fiscal year 2018, the fair competition policy criteria (C1) had the most significant impact on the increased number of anti-competition problems, as per the results of the study conducted on the European financial industry. The presence of a deficient ethical culture and the adverse consequences of successive mergers among financial institutions throughout Europe are underscored by the discovery that Policy Fair Competition outlines the ethical guidelines that financial institutions aim to adhere to in order to be fair competitors. This standard leads to an increase in anti-competition disputes. The European Organization for Competition and Consumer Protection (OECD) developed the Fair Competition Policy. In order to guarantee the financial sector's continued existence, Greece has implemented numerous bank mergers in recent years, particularly during the economic crisis. As a result of these mergers, the market became highly concentrated, as numerous companies acquired or gained control of one another through a variety of methods. The likelihood of fostering unjust rivalry among competitors and restricting competition was elevated as a result of this concentration. The Greek financial sector's banking consolidation, which resulted in the establishment of four large banks, significantly contributed to the reduction of the banking market's size. The Competition Commission was apprehensive about these mergers due to the potential for them to impede competition and exacerbate behaviors that are deemed unjust in the context of competition. To be more precise, the unethical competition strategies consisted of incidents of clandestine collaboration among executives in managerial positions at financial institutions.

Conversely, the European banking industry's anti-competition conflicts are marginally affected by the accounting challenges of the fiscal year 2018, the wage working condition disputes of the fiscal year 2019, the executive pay disputes of the fiscal years 2020–2021, and the insider deals scandals of the fiscal year 2022. Regulators enhance their oversight during crisis periods, such as the COVID-19 pandemic, in response to severe external disruptions, as demonstrated by prior research. This is in reference to the disagreement regarding the accounting criteria. Consequently, it is conceivable that bank managers may continue to experience a strong desire to influence revenue levels during periods of crisis. This is likely due to their apprehension regarding the severe repercussions that would ensue if they violated the rules, as well as their diligent efforts to achieve financial goals. As a result, they endeavor to refrain from engaging in unorthodox or oppressive accounting practices, which have the potential to lead to disputes regarding the reporting of financial information. The issue of pay and working conditions is not a significant factor in the anti-competition issues that are currently affecting financial institutions, as there is insufficient evidence to support this claim. In terms of customer service, quality, and the administration of their numerous employees, these establishments are facing intense competition.

Furthermore, recent research indicates that CEO compensation is among the least pertinent variables in anti-competition disputes, particularly during the fiscal years 2020 and 2021. This is despite the fact that executive compensation has been the subject of criticism in recent years. For the purposes of this discourse, executive pay pertains to situations in which chief executive officers depart their positions with substantial compensation, despite the fact that the organizations they oversee are experiencing financial difficulties. The media consistently reports on these incidents and presents them as evidence of an existing issue. The justification for the introduction of salary constraints on chief executive officers has been bolstered by the current financial crisis, the COVID-19 issue, and the government's involvement in the rescue of collapsing businesses using public funds. It is also feasible to contend that the government's interference enabled certain chief executive officers to resign from their positions with hundreds of millions of dollars in their wallets just as their organizations were poised to falter financially. The dot-com bubble, which emerged subsequent to the financial and real estate expansions of the preceding fifteen years, has been emphasized by a number of economists as a direct consequence of the permissive monetary policies and intentionally low-interest rates of central banks. This bubble emerged subsequent to the expansions that transpired over the subsequent 15 years. The individuals who chose to exercise their stock options at the peak of the market surges, prior to the subsequent market collapse, made the greatest profit from these irresponsible monetary policies. Additionally, a moral hazard issue substantially impeded the market's ability to self-correct in the financial sector. Investors were under the impression that the government would provide financial assistance to businesses that were perceived as being too large to fail, and this assumption was, in fact, realized in the majority of instances. This led to the dissolution of the discipline that was present within the market. This is more likely to be attributed to a government failure than to a market failure. In an effort to grant shareholders a more significant say in the determination of CEO compensation, financial organizations have implemented the "say on pay" policy. This approach provides shareholders with the opportunity to actively influence decisions regarding remuneration through the use of a vote, which may be either advisory or binding, rather than relying solely on the board of directors. Additionally, it is essential for executives to refrain from implementing strategies that could potentially result in the firm's corruption through unethical competition.

Insider trade scandals, which are also referred to as insider deal scandals, may be regarded as a severe form of corruption within an organization. It is evident that this specific component plays a substantial role in the facilitation of instances of fraud or bribery, as indicated by the results of the entropy-weighted study. A technique that prioritizes solutions based on their similarity to the ideal solution is denoted by the acronym TOPSIS. This component has the least significant anti-competitive effect. This may be due to the rigorous and dynamic corporate governance measures of European financial institutions, which significantly reduce the prevalence of insider trading or market manipulation. These policies prohibit both insider trading and market manipulation. In addition, research conducted in the disciplines of corporate governance and insider trading has underscored the possibility that the majority of insider trading incidents that occur in publicly listed companies may be attributed to insufficient corporate governance and internal control. Businesses with deficient corporate governance have a higher likelihood of engaging in insider trading. Inadequate corporate governance impedes the ability to provide clear direction and regulations for business management operations and to deliver high-quality information, whereas robust corporate governance has the potential to enhance the quality of information for publicly traded companies.

The TOPSIS methodology was employed to assess the performance of European financial institutions in relation to anti-competition issues, with the entropy-weight matrix and the decision matrix serving as inputs. The appendix contains a table that details the sub-criteria that performed the best (+) and the sub-criteria that performed the worst (−), as previously indicated. The final results are presented in a separate table. Furthermore, we present the results of the TOPSIS (technology for order preference by similarity to the ideal solution) approach, which indicate that the optimal option has the smallest disparity from the positive ideal solution and the greatest disparity from the negative ideal solution.

It is feasible to derive the conclusion that the European financial institutions with headquarters in Scandinavia have a higher than average incidence of anti-competitive disputes. This conclusion is predicated on the data presented in Table (…) and the ranking that was generated through the TOPSIS methodology. Given that the Scandinavian banking sector has been under investigation for antitrust breaches for an extended period, this disclosure may be regarded as incredible. Nevertheless, research conducted in this field has underscored the possibility that the Nordic nations may become genuine hypocrites in terms of compliance. However, what was the cause of this occurrence? Finland, Denmark, and Sweden are the nations with the highest levels of perceived corruption, as indicated by the Transparency International Corruption Perceptions Index. Denmark is presently ranked as the highest-ranked nation, with Finland and Sweden occupying the third and fourth positions, respectively. Consequently, it is probable that the inhabitants of these nations are acquainted with the principles of impartial media, democratic voting, and free expression. Their administrations' decision-making processes are generally transparent to the public, and they are not burdened by corruption on a daily basis. This is the reason for the underlying presupposition in Scandinavian culture that individuals have a propensity to conduct themselves in a moral manner.

Challenges are any occurrences that occur within the corporate environment, which emphasizes the severity of the issue. It is not surprising to individuals who are considered to be members of the compliance community that major banks in the Nordic region, including Nordea in Finland, Swedbank in Sweden, and Danske Bank in Denmark, are currently facing claims and investigations regarding significant violations of money laundering legislation. The efforts of our reputable financial institutions have returned a substantial sum of money that had been contaminated by corruption to the system. The true nature of the fundamental issue is occasionally overlooked in the midst of ongoing discussions regarding the failure of compliance systems at financial institutions. Criminals who wish to reintegrate their illicit funds into the financial system are compelled to participate in money laundering, as it is not a recreational activity but rather a necessity for them to do so efficiently. The cash in question is obtained from Nordic businesses and is used as unlawful payments to corrupt officials in authoritarian regimes in order to acquire government contracts, licenses, and corporate victories. This is done in order to capitalize on lucrative economic opportunities. These variables may result in an increase in unjust competition among financial institutions in Scandinavia as a result of the proliferation of controversial issues, such as environmental problems. This is due to the proliferation of contentious issues. In its most recent annual report, the Swedish Climate Policy Council (SCPC), issued on March 29, Sweden expressed concern regarding its development, which is experiencing a decline, while the European Union is maintaining its current trajectory and accelerating its pace. This may be a substantial factor contributing to the rise in environmental conflicts within the financial sector, which ultimately leads to the implementation of anti-competitive activities to achieve their objectives.

The findings that pertain to the nations that are situated at the bottom of the TOPSIS aggregate score list are of the utmost importance. British financial institutions are among the most problematic in terms of anti-competition law issues. The financial services industry in the United Kingdom appears to be egalitarian, at least in theory. Authorities generally lack a comprehensive understanding of the business models of firms in their respective sectors when it comes to competition. This has the potential to lead to significant deviations from the assumptions outlined in competition textbooks, particularly in the financial markets. The low rates of client attrition between providers provide additional evidence that the retail financial markets frequently lack demand-side discipline. This has a negative impact on the efficacy of numerous conventional policies that are designed to encourage competition, while it does not affect unethical competition. Nevertheless, financial regulators have discovered that the necessary level of competition to safeguard clients by influencing pricing and quality to meet their preferences is not achieved when attempting to rectify information asymmetry in situations where the demand side is insufficient. They have observed this.

5.5 Conclusions

On the front lines of the battle against fraud and corruption, resistance is motivated by a multitude of variables. In addition to eroding democratic ideals, these actions result in significant costs, which amount to a range of 179 billion to 990 billion euros annually with regard to the European Union alone. Fraud, on the other hand, weakens public trust, which is not only important but possibly far more problematic. A government's principal line of defense against fraud is established by supervisory mechanisms that are robust and effective. On the other hand, it is unfortunate that errors in these procedures might also play a role in cases of fraud. Because every effort made to address and discourage dishonest acts may accidentally create further vulnerabilities, it is a tough task to successfully identify and adequately penalize these sins. Beyond the realm of governments and the public sector, fraud is a problem that affects a wide range of organizations. There are several manifestations of it, and it may include a broad variety of individuals. It has a wide-ranging influence on a number of different enterprises. Fraud, inappropriate marketing of financial goods, manipulation of markets, conflicts of interest, insider trading, and money laundering are all examples of types of unethical behavior that fall under the umbrella of corruption and unethical behavior in the domain of finance. Furthermore, it encompasses actions that disrespect the greater responsibility that banks have toward depositors, taxpayers, and society as a whole from a broader perspective. The most recent scandals are evidence that the efforts made by regulatory and judicial organizations have not been successful in bringing about the indispensable and necessary cultural and structural changes that are required to reclaim people's confidence.

In the European Union, the regulations governing business integrity have not yet reached a level that is enough to effectively prohibit corrupt practices and undue influence on businesses. Both the causes and the impacts of corruption are not effectively addressed by the governance structure that the European Union has established for commercial activities. For the purpose of this study, a limited body of research and an in-depth investigation into the elements that greatly impact scenarios that might lead to the worsening of corruption issues, such as the emergence of anti-competition conflicts, have been investigated. Anti-environmental, social, and governance (ESG) components, which are strongly related to fraudulent activities and corrupt practices, are the components that merit inquiry and have the potential to develop corruption inside an organization. As a result, the current research is a groundbreaking endeavor that uses the TOPSIS approach to investigate and evaluate the degree to which anti-ESG factors impact anti-competition conflicts in the European banking sector. For the purpose of determining the magnitude of each of the selected anti-ESG components, the entropy-weighted TOPSIS approach was also used.

Examining the effectiveness of anti-competition conflicts in European financial institutions is the primary focus of the conclusions of the present research. In comparison to the rest of Europe, Scandinavian financial institutions have shown superior performance in terms of anti-competition tactics. This is one of the most important findings. There is a consensus among experts that Scandinavia is among the most forward-thinking areas in the world. In a situation that is rather amazing to see, banks in Scandinavia are presently in the midst of a crisis that is associated with anti-competitive activity and money laundering. In the name of Russian billionaires, corrupt politicians, and leaders of organized crime, these financial institutions are being accused of aiding the movement of considerable quantities of illegal cash to offshore tax havens. The amount of these monies is estimated to be in the hundreds of billions of dollars. The United Kingdom, on the other hand, is regarded as one of the nations that performs better in mitigating anti-competition concerns in the financial industry. This is because the United Kingdom is the largest financial center in Central Europe. The rationale for this is that financial regulators have discovered that resolving information asymmetry in situations where the demand side is weak does not result in the needed amount of competition that safeguards customers by influencing pricing and quality to match with their preferences. This is the basis for this conclusion.

5.5.1 Implications and Future Lines

The outcomes of our study not only highlight significant policy consequences, but they also give theoretical foundations and understanding for results that have been experimentally proven. This study is the first of its kind to place an emphasis on the concept of anti-competition conflicts and the role that these disputes play in the development of corruption, bribery, and fraud within the African banking sector. Although anti-competitive tactics and corrupt practices may seem to be two separate concepts, they are quite similar in that they both lead to the financial sector having an excessive amount of power over the market. The result of this is that there is an increase in expenses and inefficiencies, which eventually has an effect on the pricing of financial products and services as well as their overall quality. The failure to acquire the greatest value for money might be the consequence of these two breaches, which can have complementary impacts. This failure can occur when contracts or markets are awarded based on reasons other than fair competition and the contractor's expertise. Furthermore, bribery and corrupt activities are often related to acts of co-operation and behaviors that restrict competition. Some examples of these behaviors are price manipulation, market division, and bid rigging. As a consequence of this, various investigations that are being conducted concentrate on allegations of antitrust violations and corruption. A preliminary inquiry into the links between the three ESG (environmental, social, and governance) components and their influence on bank corruption is based on the incidence of anti-competition events, which serves as the foundation for this study's first examination. Offering insights into the identification of the most and least corrupt financial institutions in Europe is the objective of our study, which aims to fill a gap in the existing body of information by offering expertise. The present research also identifies the factors that the market has to take into consideration when formulating an anti-corruption strategy that is successful. Using multiple-criteria decision analysis (MCDM), we have constructed two models in order to assess the factors. The first model used the TOPSIS approach, while the second model included the entropy weight. Both models were developed in order to analyze the elements. It is a TOPSIS.

Potentially informing CEOs on how to build successful anti-corruption measures is one of the practical results that might be derived from the present study. It is the primary emphasis of financial institutions to examine the many financial factors that have the potential to lead to corruption inside the company. Furthermore, financial institutions have not yet performed comprehensive investigations into organizations that have close relationships with issues, which in turn has the potential to lead to cases of fraud and bribery having occurred. We thus give, for the first time, to the governing bodies of European financial institutions the particular aspects that are connected with anti-ESG actions and have the potential to generate corruption. This is how our current research delivers the information. In light of this, financial institutions are able to create appropriate anti-corruption measures by putting an emphasis on these fundamental characteristics. These measures may help eliminate corruption and the phenomena that are linked with it, such as unfair competition. One kind of competition in particular has the potential to have a direct influence on both the internal and external environments of businesses that are active in the financial industry. Financial institutions are able to get a thorough awareness of the vital factors that they should integrate into their anti-corruption activities if they are able to elucidate the qualities that might magnify inequitable competition and are related to elements that are anti-environmental, social, and governance policies.

Additionally, the investigation of the same criteria from a digital finance perspective might be prioritized in the research that will be conducted in the future. Over the last several years, there has been a significant increase in the use of digital finance, which provides clients with quick access to their accounts and enables financial institutions to improve their systems. In spite of this, the introduction of digital banking has resulted in an increased vulnerability to fraudulent actions taking place. For financial institutions and their customers, the identification and prevention of fraudulent behavior is becoming an increasingly difficult task as con artists get more skilled at the tactics they use to commit fraud. An evaluation of the performance of companies operating in the digital finance industry may thus be accomplished via the use of a combination of Multiple-Criteria Decision Making (MCDM) techniques. The acronyms TOPSIS, Critic-TOPSIS, Analytic Hierarchy Process (AHP), and ELimination Et Choix Traduisant la R'Ealit'E (ELECTRE) are included in this category. It is vital for people to have an understanding of these issues in order to protect themselves against financial losses, corruption, bribery, and fraud. Proactive measures may be implemented by financial institutions and banks in order to safeguard the assets and information of their customers.

References

Agnese, P., Battaglia, F., Busato, F., & Taddeo, S. (2023). ESG controversies and governance: Evidence from the banking industry. *Finance Research Letters, 53*. https://doi.org/10.1016/j.frl.2022.103397

Aouadi, A., & Marsat, S. (2018, September). Do ESG controversies matter for firm value? Evidence from international data. *Journal of Business Ethics*, Springer, *151*(4), 1027–1047.

Al-Sadig, A. (2009). The effects of corruption on FDI inflows. *Cato Journal, 29*(2), 267–294. https://EconPapers.repec.org/RePEc:cto:journl:v:29:y:2009:i:2:p:267-294

Amankwaah, E., & Baidoo, N. O. (2023). Effect of firm size and corruption on financial challenges of savings and loans companies: Evidence from Ghana. *Social Sciences and Humanities Open, 8*(1). https://doi.org/10.1016/j.ssaho.2023.100552

Anita, M., Shveta, S., Yadav Surendra, S., & Arvind, M. (2023). When do ESG controversies reduce firm value in India? *Global Finance Journal, 55*. https://doi.org/10.1016/j.gfj.2023.100809

Atenidegbe, O. F., & Mogaji, K. A. (2023). Modeling assessment of groundwater vulnerability to contamination risk in a typical basement terrain using TOPSIS-entropy developed vulnerability data mining technique. *Heliyon, 9*(7). https://doi.org/10.1016/j.heliyon.2023.e18371

Balsalobre-Lorente, D., Topaloglu, E. E., Nur, T., & Evcimen, C. (2023). Exploring the linkage between financial development and ecological footprint in APEC countries: A novel view under corruption perception and environmental policy stringency. *Journal of Cleaner Production, 414*. https://doi.org/10.1016/j.jclepro.2023.137686

Bang, J., Ryu, D., & Yu, J. (2023). ESG controversies and investor trading behavior in the Korean market. *Finance Research Letters, 54*. https://doi.org/10.1016/j.frl.2023.103750

Basnyat, B., Treue, T., Pokharel, R. K., Kayastha, P. K., & Shrestha, G. K. (2023). Conservation by corruption: The hidden yet regulated economy in Nepal's community forest timber sector. *Forest Policy and Economics, 149*. https://doi.org/10.1016/j.forpol.2023.102917

Cernohorska, L. (2015). Impact of financial crisis on the stability banking sectors in The Czech Republic and Great Britain. *Procedia Economics and Finance, 26*, 234–241. https://doi.org/10.1016/s2212-5671(15)00824-2

Chu, H., Sun, S., & Wei, J. (2023). Fiscal pressure and judicial decisions: Evidence from financial penalties for official corruption in China. *International Review of Law and Economics*, 106156. https://doi.org/10.1016/j.irle.2023.106156

Cicchiello, A. F., Cotugno, M., & Foroni, C. (2023). Does competition affect ESG controversies? Evidence from the banking industry. *Finance Research Letters, 55*. https://doi.org/10.1016/j.frl.2023.103972

Cuervo-Cazurra, A. (2006). Who cares about corruption? *Journal of International Business Studies, 37*(6), 807–822. https://doi.org/10.1057/PALGRAVE.JIBS.8400223

DasGupta, R. (2022). Financial performance shortfall, ESG controversies, and ESG performance: Evidence from firms around the world. *Finance Research Letters, 46*. https://doi.org/10.1016/j.frl.2021.102487

Dwivedi, P. P., & Sharma, D. K. (2023). Evaluation and ranking of battery electric vehicles by Shannon's entropy and TOPSIS methods. *Mathematics and Computers in Simulation, 212*, 457–474. https://doi.org/10.1016/j.matcom.2023.05.013

Hwang, C.-L., & Yoon, K. (1981). *Multiple Attribute Decision Making, 186*. https://doi.org/10.1007/978-3-642-48318-9

Kaur, H., Gupta, S., & Dhingra, A. (2023). Selection of solar panel using entropy TOPSIS technique. *Materials Today: Proceedings*. https://doi.org/10.1016/j.matpr.2023.02.034

Khan, M. A., Gu, L., Khan, M. A., & Bhatti, M. I. (2022). Institutional perspective of financial sector development: A multidimensional assessment. *Economic Systems, 46*(4). https://doi.org/10.1016/j.ecosys.2022.101041

Knack, S., & Keefer, P. (1997). Does social capital have an economic payoff? A cross-country investigation. *The Quarterly Journal of Economics, 112*(4), 1251–1288. https://doi.org/10.1162/003355300555475

Latapí, M., Davíðsdóttir, B., & Jóhannsdóttir, L. (2023). Drivers and barriers for the large-scale adoption of hydrogen fuel cells by Nordic shipping companies. *International Journal of Hydrogen Energy, 48*(15), 6099–6119. https://doi.org/10.1016/j.ijhydene.2022.11.108

Lindman, Å., Ranängen, H., & Kauppila, O. (2020). Guiding corporate social responsibility practice for social license to operate: A Nordic mining perspective. *Extractive Industries and Society, 7*(3), 892–907. https://doi.org/10.1016/j.exis.2020.07.013

Liu, Q., Ma, J., Zhao, X., Zhang, K., Xiangli, K., Meng, D., & Wang, J. (2023a). Voltage fault diagnosis and misdiagnosis analysis of battery systems using the modified Shannon entropy in real-world electric vehicles. *Journal of Energy Storage, 73*. https://doi.org/10.1016/j.est.2023.109287

Liu, J., Qian, H., Zhang, Q., Lin, Z., & Siano, P. (2023b). Corruption induced energy inefficiencies: Evidence from China's energy investment projects. *Energy Policy, 183*. https://doi.org/10.1016/j.enpol.2023.113825

Li, X., Lin, C., & Zhan, X. (2018). Does change in the information environment affect financing choices? *Management Science, 65*(12), 5676–5696. https://doi.org/10.1287/mnsc.2018.3096

Lucey, B. M., Kumar, S., & Sureka, R. (2023). Corruption in finance research: The state of art and future research agenda. *Journal of Economic Criminology, 1*, 100001. https://doi.org/10.1016/j.jeconc.2023.100001

Marsat S., Pijourlet, G., Ullah, M., (2022). Does environmental performance help firms to be more resilient against environmental controversies? International evidence. *Finance Research Letters, 44*, 102028. ISSN 1544-6123. https://doi.org/10.1016/j.frl.2021.102028

Priya, P., & Sharma, C. (2023). Reinforcing the effects of corruption and financial constraints on firm performance: Normal versus crisis period in developing economies. *Economic Modelling, 127*. https://doi.org/10.1016/j.econmod.2023.106463

Roy, P., Watkins, M., Iwuamadi, C. K., & Ibrahim, J. (2023). Breaking the cycle of corruption in Nigeria's electricity sector: Off-grid solutions for local enterprises. *Energy Research and Social Science, 101*. https://doi.org/10.1016/j.erss.2023.103130

Saraiva, P. (2023). On Shannon entropy and its applications. *Kuwait Journal of Science, 50*(3), 194–199. https://doi.org/10.1016/j.kjs.2023.05.004

Shakil, M. H. (2021). Environmental, social and governance performance and financial risk: Moderating role of ESG controversies and board gender diversity. *Resources Policy, 72*. https://doi.org/10.1016/j.resourpol.2021.102144

Su, F., & Xu, C. (2023). Curbing credit corruption in China: The role of FinTech. *Journal of Innovation and Knowledge, 8*(1). https://doi.org/10.1016/j.jik.2022.100292

Sun, S. L., Ding, Z., & Joseph, G. (2023). Expanding inclusive markets through corruption control: A multilevel modeling analysis for a grand challenge. *Journal of International Management*. https://doi.org/10.1016/j.intman.2023.101068

Tchapchet Tchouto, J.-E. (2023). An empirical assessment on the leveraging evidence of economic complexity under environmental Kuznets curve hypothesis: A comparative analysis between Nordic and non-Nordic European countries. *Innovation and Green Development, 2*(4), 100074. https://doi.org/10.1016/j.igd.2023.100074

Zare Banadkouki, M. R. (2023). Selection of strategies to improve energy efficiency in industry: A hybrid approach using entropy weight method and fuzzy TOPSIS. *Energy, 279*. https://doi.org/10.1016/j.energy.2023.128070

The manufacturer's authorised representative in the EU is Springer Nature Customer Service Centre GmbH, Europaplatz 3, 69115 Heidelberg, Germany. If you have any concerns regarding our products, please contact ProductSafety@springernature.com

Printed and bound by CPI Group (UK) Ltd, Croydon, CR0 4YY
05/10/2026
02228156-0001